Measuring Behaviour
An Introductory Guide

This third edition of *Measuring Behaviour* has been largely rewritten and reorgan-ised. As before, however, it is a guide to the principles and methods of quantitative studies of behaviour, with an emphasis on techniques of observation, recording and analysis. It provides the basic knowledge needed to measure behaviour, doing so in a succinct and easily understood form.

Aimed primarily at undergraduate and graduate students in biology and psy-chology who are about to embark upon quantitative studies of animal and human behaviour, this book provides a concise review of methodology that will be of great value to scientists of all disciplines in which behaviour is measured, includ-ing the social sciences and medicine. Principles and techniques are explained clearly in simple and concise language. Its most general points apply to many other biological sciences.

Measuring Behaviour has established itself as a standard text in its field. This third edition has been completely updated. The sections on research design and on the interpretation and presentation of data have been greatly expanded. Written with brevity and clarity, *Measuring Behaviour* is intended, above all, as a practical guide book.

DR PAUL MARTIN studied biology at Cambridge University, graduating in natural sciences and with a Ph.D. in behavioural biology. A former Harkness Fellow in the Department of Psychiatry & Behavioural Sciences at Stanford University in the USA, he has lectured and researched in behavioural biology at Cambridge, and was Fellow of Wolfson College, Cambridge.

PROFESSOR SIR PATRICK BATESON has been a highly esteemed lecturer in animal behaviour across the world, for nearly 40 years. He is a Fellow of the Royal Society, a former President of the Association for the Study of Animal Behaviour, Knight Batchelor and current president of the Zoological Society of London.

Measuring Behaviour

An Introductory Guide

THIRD EDITION

Paul Martin
Former University Demonstrator in Animal Behaviour at the University of Cambridge
Sub-Department of Animal Behaviour and Fellow of Wolfson College, Cambridge

Patrick Bateson
Emeritus Professor of Ethology at the University of Cambridge, Fellow of King's College,
Cambridge and President of the Zoological Society of London

CAMBRIDGE
UNIVERSITY PRESS

CAMBRIDGE UNIVERSITY PRESS

Cambridge, New York, Melbourne, Madrid, Cape Town, Singapore, São Paulo,
Delhi, Tokyo, Mexico City

Cambridge University Press
The Edinburgh Building, Cambridge CB2 8RU, UK

Published in the United States of America by Cambridge University Press, New York

www.cambridge.org
Information on this title: www.cambridge.org/9780521828680

First published 2007
Sixth printing 2011

Printed in the United Kingdom at the University Press, Cambridge

A catalogue record for this publication is available from the British Library

ISBN 978-0-521-82868-0 Hardback
ISBN 978-0-521-53563-2 Paperback

CONTENTS

PREFACE

We are pleased that many of the issues that were relatively novel in behavioural biology when we wrote the first edition (1986) of this book have now passed into the mainstream of methodological thought. Nevertheless, we believe that the principles are worth reinforcing.

In this edition we have changed the structure so that greater prominence is given to the non-experimental aspects of behavioural biology. Some behavioural research simply involves carefully watching an animal to see what it does next. Performing an experiment may seem more 'scientific' than open-ended observation but the yield may be less. Moreover, worthwhile experimental research almost invariably needs to be preceded by careful observation. Knowledge of the normal behaviour of animals, preferably in their natural environment, is an invaluable precursor to experimental research.

We have also expanded the section on research design because, more than ever, good design can make such a difference to how big the sample must be, the interpretation of data and the time taken to prepare results for presentation or publication when the moment arrives. We have eliminated the further reading sections at the end of each chapter, but have given advice on further reading at appropriate places in the chapters. Each chapter now ends with a summary. We have taken out the annotated bibliography that formed such a large part of the reference section in the two previous editions (1986 and 1993) because we felt that such material was not essential to the main purpose of the book. However, we have included some advice on statistics books in Appendix 3 and, since it contains many references to important papers on the methodology of measuring and analysing behaviour, we have put the annotated bibliography of the second edition on the following website: www.cus.cam.ac.uk/~ppgb/.

The second edition of this book was published in 1993 and many things have changed since then. Technology moves particularly rapidly. For that reason we have reduced the amount of material that is likely to pass out of date quickly and suggest to those planning a new project that you keep yourselves abreast of new technological developments. Much the same is true for statistical techniques where changes are also taking place. We have not tried to make this book so self-contained that no other source is needed. The underlying principles of measuring behaviour, however, do not change rapidly. We have focused on these principles in this new edition, as we did in the first two.

As before, we have given relatively few examples in the text. To do so would have made it much longer. We resolved to keep the book slim and reasonably priced. Above all we wanted it to remain what it always has been – an introductory guide to the basic problems and possibilities of measuring behaviour. For those who want more, the second edition of Philip Lehner's (1996) *Handbook of Ethological Methods* is filled with excellent examples, as is J. D. Paterson's (2001) second edition of *Primate Behavior*, aimed at primatologists working on behaviour.

Inevitably in such a short book, we have dealt with many complex and contentious issues rather briefly, and in some cases the advice we offer is based on opinions that are not universally shared. We hope you will note our cautions and, where necessary, explore the issues in greater depth than is possible here. We also hope that, if you are coming to this book for the first time, you will read the first two chapters. These chapters set the scene for anybody proposing to start research involving the measurement of behaviour of animals – including the human animal.

In preparing this new edition we were helped by our researchers Elizabeth Pimley and Guy Martin and we thank them for their work. Thanks to the efforts of Tracey Sanderson, our Cambridge University Press editor at the time, 20 anonymous referees provided us with comments on the second edition and made many helpful suggestions for improvements in the third. Three graduate students, Chris Bird, Anne Helme and Amanda Seed, read through the full text of the new edition with fresh eyes, and gave us much encouragement and good advice. We also received expert help on particular issues from Alan Grafen, Hanna Kokko, Peter Lipton and Marc Mangel. Finally, Martin Griffiths, our current Cambridge University Press editor, has been enthusiastic about what has become a largely new project and generously accepted the long delays in preparing this third edition. To all of these people we are greatly indebted.

1

Introduction

The scope of this book

This book is intended as a guide to all those who are about to start work involving the measurement of directly observed behaviour. We hope it will also be useful for those wanting to refresh their memories about both the possibilities and the shortcomings of available techniques.

Those who have never attempted to measure behaviour may suppose from the safety of an armchair that the job is an easy and straightforward one, requiring no special knowledge or skills. Is it not simply a matter of writing down what happens? In sharp contrast, those attempting to make systematic measurements of behaviour for the first time are often appalled by the apparent difficulty of the job facing them. How will they ever notice, let alone record accurately and systematically, all that is happening? The truth is that measuring behaviour *is* a skill, but not one that is especially difficult to master, given some basic knowledge and an awareness of the possible pitfalls.

The purpose of this book is to provide the basic knowledge in a succinct and easily understood form, enabling the beginner to start measuring behaviour accurately and reliably. A great deal of high-quality behavioural research can be done without the need for specialised skills or elaborate and expensive equipment.

Sometimes it is possible to carry out behavioural research simply by relying on written descriptions of what the subjects do. Usually, though, worthwhile research will require that at least some aspects of the behaviour are measured. By *measure* we mean quantify by assigning numbers to observations according to specified rules. Therefore, measurement

Table 1.1 *The four problems of behavioural biology.*

	Current	Historical
Proximate:	How does it work?	How did it develop?
Ultimate:	What is it for?	How did it evolve?

The control and development questions are sometimes grouped as proximate, and the functional and evolutionary questions as ultimate. They may also be grouped as historical and current issues to do with fully formed behaviour. From Tinbergen, 1963.

of behaviour, whether in the laboratory or in the field, is required by virtually all behavioural biologists and psychologists.

In this book, we are primarily concerned with the methods based on *direct observation* of behaviour developed for recording the activities of non-human species. These methods are not only applicable in advanced academic research. They may also be readily used in the behavioural projects that are commonly offered in university teaching courses and in some secondary schools. Moreover, even though the techniques do not deal with some important issues, such as the measurement and analysis of language, they may be applied fruitfully in some studies of human behaviour, and therefore have important uses in the social and medical sciences.

The four problems

A number of fundamentally different types of question may be asked when studying behaviour. Probably the most useful and widely accepted classification was formulated by the Nobel prize-winning ethologist, Niko Tinbergen (1963), who pointed out that four distinct types of problem are raised by the study of behaviour (see Table 1.1).

Proximate causation or *control* – 'How does it work?' How do internal and external causal factors elicit and control behaviour in the short term? For example, which stimuli elicit the behaviour pattern and what are the underlying neurobiological, psychological or physiological mechanisms regulating the animal's behaviour?

Development or *ontogeny* – 'How did it develop?' How did the behaviour arise during the lifetime of the individual; that is, how is behaviour assembled? What factors influence the way in which behaviour develops during the lifetime of the individual and how do the developmental processes work? What is the interplay between the individual and its environment during the assembly of its behaviour? In addition, what aspects of the young animal's behaviour are specialisations for dealing with the problems of early life or the gathering of information required for behavioural development? Answering these questions is the behavioural side of developmental biology.

Function – 'What is it for?' What is the current use or survival value of the behaviour? How does behaving in a particular way help the individual to survive? How does its behaviour help it to reproduce in its physical and social environment?

Evolution or *phylogeny* – 'How did it evolve?' How did the behaviour arise during the evolutionary history of the species? What factors might have been involved in moulding the behaviour over the course of evolutionary history? How can comparisons between different species help to explain that history? How has behaviour itself driven the evolutionary process through mate choice and animals' adaptability and construction of their environments?

Evolutionary questions are concerned with the historical origins of behaviour patterns, whereas functional questions concern their current utility. The two are frequently confused. Questions of function and evolution are sometimes referred to as 'ultimate' questions when contrasted with proximate causation. In the case of fully formed behaviour, questions to do with control and function are current, whereas questions to do with evolution and development are historical.

The 'Four Problems' can perhaps best be illustrated with a commonplace example. Suppose we ask why it is that drivers stop their cars at red traffic lights. One answer would be that a specific visual stimulus – the red light – is perceived, processed in the central nervous system and reliably elicits a specific response (easing off on the accelerator, applying the brake and so on). This would be an explanation in terms of proximate causation. A different but equally valid answer is that individual drivers have learnt this rule by past observation and instruction. This is an explanation in terms of development. A functional explanation is that drivers who do not stop at red traffic lights are liable to have an accident or, at least, be stopped by the police. Finally, an 'evolutionary' explanation would deal with the historical processes whereby a red light came to be used as a universal signal for stopping traffic at road junctions. All four answers are equally correct, but reflect four distinct levels of enquiry about the same phenomenon.

While Tinbergen's Four Problems are logically distinct and should not be confused with each other, it can be helpful to ask more than one type of question at the same time. Correlations between the occurrence of behaviour and the circumstances in which it is seen often lead to speculations about current function. These speculations can lead in two directions. They may suggest what are likely to be important controlling variables and thence lead to experiment. Alternatively, they may suggest a design for the way in which the mechanism ought to work. Here again, the proposal can be tested against reality. For example, as an animal gathers information about its fluctuating environment, what rules should it use in deciding where it should feed? Should it go to a place where the food is always available in small amounts or one in which it is periodically available in large amounts? Ideas about the best ways to sort out such conflicts between foraging in different places have provided insights into the nature of the mechanism. Working the other way, knowledge of mechanism has provided understanding of how such behaviour might have evolved (Real, 1994).

Different approaches to studying behaviour

Scientists study behaviour in different ways and for very different reasons. Historically, **psychology** (which originally grew out of the study of the human mind) was distinguished from **ethology** (the biological study of behaviour) in terms of the methods, interests and the origins of the two sciences. In the twentieth century, comparative and experimental psychologists tended to focus mainly on questions about the proximate causation of behaviour (so-called 'how' questions), studying general processes of behaviour (notably learning) in a few species under laboratory conditions. In contrast, ethologists had their roots in biology and asked questions not only about how behaviour is controlled but also about what behaviour is for and how it evolved ('why' questions). A superb account of the origins of modern ethology is given by Burkhardt (2005).

Biologists have been trained to compare and contrast diverse species. Impregnated as their thinking is with the Darwinian theory of evolution, they habitually think about the adaptive significance of the differences between and within species. Indeed, many ethologists were primarily interested in the biological utility of behaviour and were wary of proceeding far with laboratory experiments without first understanding the function of the behaviour in its natural context. Studies in unconstrained conditions of animals, including humans, have been an important feature of ethology and have played a major role in developing the distinctive and powerful methods for observing and measuring behaviour. In contrast, psychologists traditionally placed greater emphasis on experimental design and quantitative methods.

Even so, it would be a fundamental mistake to represent modern ethology as non-experimental or psychology as non-observational. A great many people who call themselves ethologists have devoted much of their professional lives to laboratory studies of the control and development of behaviour. Conversely, many people who call themselves psychologists study their subjects in unconstrained environments.

Field studies that related behaviour patterns to the social and ecological conditions in which they normally occur led to the development of behavioural ecology. Another sub-discipline, sociobiology, brought to the study of behaviour important concepts and methods from population biology and stimulated further interest in field studies of animal behaviour. Sociobiology also spawned the subject of evolutionary psychology, which seeks to apply biological principles of evolution to human behaviour in particular (Barrett *et al.*, 2002). These subjects have sometimes ignored issues to do with the workings of behaviour. Gradually, however, it became apparent that such neglect of important areas of the biology of behaviour was a mistake. An understanding of the development and control of behaviour is important in stimulating (as well as constraining) ideas about function and evolution. The stimulus works both ways and all the sub-disciplines have started to merge.

For simplicity, we shall refer henceforth to ethology, behavioural ecology, sociobiology and the biological aspects of evolutionary psychology collectively as behavioural biology. Modern behavioural biology abuts many different disciplines and defies simple definition in terms of a common problem or shared ideas. The methods developed for the measurement of behaviour are used by neurobiologists, behaviour geneticists, social and developmental psychologists, anthropologists and psychiatrists, among many others. Considerable transfer of ideas and a convergence of thinking have occurred between behavioural biology and psychology, from which both subjects have greatly benefited. For example, experimental methods developed by psychologists are being used by behavioural ecologists interested in how animals forage for food. Conversely, many of the observational methods developed by behavioural biologists have proved highly effective in studying the developmental psychology of children.

Why measure behaviour?

Animals use their freedom to move and interact, both with their environment and with one another, as one of the most important ways in which they adapt themselves to the conditions in which they live. These adaptations take many different forms such as finding food, avoiding being eaten, finding a suitable place to live, attracting a mate and caring for young. Each species has special requirements, and the same problem is often solved in different ways by different species.

Even though much is already known about such adaptations and the ways in which they are refined, as individuals gather experience, a great deal remains to be discovered about the diversity and functions of behaviour. The principles involved in the evolution of increasingly complex behaviour and the role that behaviour itself has played in shaping the direction of evolution are still not well understood. Explanations of how behaviour patterns have arisen and what they are for will only come from the comparative study of different species and by relating behaviour to the social and ecological conditions in which an animal lives.

Animals are studied for many reasons. In addition to its intrinsic interest (and, let us be frank, the fun of it), the study of behaviour is both intellectually challenging and practically important. Medical research, to which behavioural work often contributes, aims for outcomes that are of direct benefit to humans. Such work on animals is justified in the eyes of most members of the public, its usefulness having been vindicated by independent assessments (House of Lords, 2002; Nuffield, 2004). However, much research on animals is not directed primarily towards producing human benefit, even if this may be a longer-term consequence. While it may play a role in the conservation of the animal in question, first and foremost it is aimed at the fundamental understanding of biology. Animals such as rats or pigeons may be used for studying general processes of associative learning. Other animals may be studied because they are particularly appropriate subjects in which to investigate an important or remarkable phenomenon such as song development in birds or communication in honey bees.

What about mechanism? Molecular and cellular approaches to biology have made remarkable progress in the last 50 years. The neurosciences have been uncovering how nervous systems work, and the long-standing goal of understanding behaviour in terms of underlying processes is becoming attainable at last. The neurophysiological, biochemical and hormonal mechanisms underlying a number of relatively simple behaviour patterns have been uncovered in a variety of species. Powerful techniques are enabling neuroscientists to analyse the electrical and chemical functioning of specific regions of the conscious human brain. Many behavioural biologists use molecular techniques (such as the measurement of satellite DNA) to establish how closely related are the individuals in their study group.

So why bother with the measurement of *behaviour* – meaning the actions and reactions of whole organisms – when you could instead look at its underlying mechanisms? The answer is illustrated, in part, by a simple analogy. Perfect knowledge of how many times each letter of the alphabet occurs on this page would give no indication of the text's meaning. The letters must be formed into words and the words into sentences

Figure 1.1 A jumble of letters at one level forms an understandable phrase at a higher level of organisation, illustrating the point that detailed knowledge about behaviour at, say, the physiological level would not be helpful in making sense of what was happening at the behavioural level.

(see Fig. 1.1). Each successive level of organisation has properties that cannot be predicted from knowing the lower levels of organisation.

Thus, even when the understanding of the neural elements underlying behaviour is complete, it will not be possible to predict how they perform as a whole without first understanding what they do as a whole – and that means knowing how the whole organism behaves. Many neuroscientists and molecular biologists are beginning to appreciate that an understanding of the mechanisms underlying behaviour requires more than exquisite analytical techniques: it also requires an understanding of the behaviour itself.

Perhaps most satisfying of all, therefore, are the studies in which an analysis of an organism's behaviour is closely integrated with an analysis of the neural, physiological and molecular mechanisms that underlie its actions. Knowledge of mechanism can greatly inform understanding of behaviour – and vice versa. Behavioural studies of imprinting in birds led to extensive analyses of the neural basis of the recognition process. This work then raised questions about the role of the neural mechanisms involved in the classical and operant conditioning that occur in parallel with imprinting. Attention was drawn back to what happens at the behavioural level (Bateson, 2005b).

For these and other reasons we believe that an essential part of biology will be the thorough description and analysis of behaviour. We hope that this book will enable you to do just that.

Summary

The direct observation of behaviour is used for many purposes, ranging from understanding cognition and how it develops to investigations of the current utility of behaviour and how it evolved in the past. The techniques for measuring behaviour are being combined with molecular and physiological approaches, but their use is important in its own right and will always remain a central part of behavioural biology and psychology.

2

Think before you measure

Many students are given ready-made problems on which to work but it pays to think carefully before you start a project, whatever stage you are at in your scientific career. Sage advice is given in the book by Cohen and Medley (2000). Here we are concerned with the particular issues that need prior thought in behavioural biology and psychology.

Choosing the level of analysis

Behaviour can be analysed at many different levels, from the complex social interactions within populations to the fine spatial detail of an individual organism's movements. A simple but fundamental point is that the form of measurement used for studying behaviour should reflect the nature of the problem and the questions posed. Conversely, the sorts of phenomena that are uncovered by a behavioural study will inevitably reflect the methods used.

A fine-grained analysis is only appropriate for answering some sorts of question, and a full understanding will not necessarily emerge from describing and analysing behaviour at the most detailed level. While a microscope is an invaluable tool, in some circumstances it would be useless – say, for reading a novel. In other words, the cost of gaining detail can be that higher-level patterns, which may be the most important or relevant features, are lost from view. For example, recording the precise three-dimensional pattern of movements for each limb may be desirable for certain purposes, such as analysing the neurophysiological mechanisms underlying a particular locomotor behaviour pattern. However, higher-level categories such as 'walk' or 'run' are often more appropriate.

On a more practical note, recording large amounts of unnecessary detail may obscure broader issues, simply by presenting you with overwhelming quantities of data to analyse and interpret.

The pattern of events you notice is affected by spatial scale: being close to a subject reveals details that would not be seen at a distance; being at a distance may reveal the wider context in which the behaviour is being expressed. Imagine that you are an observer among a group of migrating wildebeest on the Serengeti plain of East Africa. The impressions you would form close up to the animals would be totally different from those obtained from an aeroplane, where you would see the long columns of animals. Similar arguments apply to the time scale of measurements. For example, the value of 'freezing' time was demonstrated when the first photographs of galloping horses were taken in the nineteenth century. Before the invention of photography, artists had conventionally painted galloping horses as having their front and back legs extended simultaneously: photographs revealed that a horse never does this.

Choosing the species

For many researchers, choosing which species to study is not an issue. They may be most interested in and only want to study humans; they may study an animal because only a handful of that particular species is left on the planet; they may have no option other than to work on the species that is bred in a particular laboratory; and so forth. Nonetheless, when choice is possible, careful thought should be given to the attractions and problems presented by the vast number of species available to study. The wealth and diversity of zoological material is so great that time invested in finding a species that is suitable for the problem to be investigated is likely to be amply repaid later in the study. Table 2.1 lists points that are worth considering.

Choosing where to study

A basic aim of scientific research is to help distinguish between competing hypotheses and thereby to reduce the number of different ways in

Table 2.1 *Questions to consider when choosing a species of animal on which to work.*

1. Is the species easily seen in its natural habitat or readily available for study in captive conditions?
2. If it has to be imported, are the conditions for collecting it in the country of origin ethical?
3. Will long delays ensue because of quarantine requirements?
4. Is it tolerant of human presence?
5. Does it handle well if it is to be kept in captivity or hand-reared?
6. Does it breed successfully in captivity?
7. If it is to be kept in captivity, does it have any problematic dietary requirements?
8. Will feeding and housing be financially feasible if large numbers are required?
9. Will adequate veterinary care be available?
10. Are animals that are free of disease readily available from suppliers?
11. What are its life-history characteristics such as gestation period, age of independence and age of sexual maturity?
12. Is its life-span long enough to make repeated measurements possible?
13. Is its life-span short enough to make studies of development practicable?
14. Is much known about its natural history, anatomy and physiology?
15. Does an extensive biological and behavioural literature exist for this species?
16. Would controlled programmes of breeding be possible?
17. If its genetics are well described, are inbred strains available, so that variability in behaviour can be reduced?
18. Is much known about its behaviour?
19. At what time of day is it typically active?
20. How solitary or gregarious is it?
21. Is its natural behaviour suited to the particular problem which is to be investigated?
22. Does it move slowly enough to make observation relatively easy, yet fast enough to make observation rewarding?
23. If a general problem relevant to many species is to be studied, is the particular species chosen suitably representative?
24. If the species is to be studied for its own sake, can knowledge of it contribute to an understanding of human behaviour?
25. Are there opportunities for comparing its behaviour with that of other closely related species?

which the natural world can be explained. In this respect, a distinction between *experimental* research (in which conditions are actively manipulated) and purely *observational* research is not fundamental, since both can generate empirical data that distinguish between competing hypotheses. Remember that in some areas of science, such as astronomy and geology, conventional experiments are rarely possible, yet detailed quantitative hypotheses are regularly formulated and tested by observation.

Working in the controlled conditions of a laboratory on animals that have known histories has obvious benefits. Nevertheless, some of the best and most influential studies in behavioural biology have been of free-living animals. A captive animal is usually too constrained by its artificial environment to perform even a small fraction of the activities of which it is capable. Furthermore, experimental evidence that a particular factor *can* influence behaviour may not mean that it *does* influence the behaviour of free-living individuals.

To observe the full richness of an individual's behavioural repertoire and understand the conditions to which each activity is adapted, the species must usually be studied in the 'field' – in the broad sense of any environment in which individuals can range freely and interact with their own and other species. The observer notices the circumstances in which an activity is performed and those in which it never occurs, thereby obtaining clues as to what the behaviour pattern might be for (its function) and how it is controlled (its proximate causation). Useful evidence also comes from comparisons between different species of free-living animals.

A major justification for fieldwork is that it uncovers aspects of behaviour that would not otherwise be known about, providing the raw material from which research questions and hypotheses can be formulated. Moreover, it provides an understanding of how an animal's behaviour is adapted to the environment in which it normally lives, in the same way that its anatomical or physiological characteristics are adapted. Field studies have been particularly valuable in relating behaviour patterns to the social and ecological conditions in which the animals normally live. Studies in unconstrained conditions have therefore been an important feature of behavioural biology.

The need for field studies is clear, but the practical and other difficulties inherent in such work should not be under-estimated. An animal under observation may frequently disappear from view, wrecking the best-laid plans for systematic recording over a fixed period. Similarly, ensuring randomness in the choice of individuals to be studied within a population may not be easy because some identified individuals are difficult to find (see 'sampling rules' in Chapter 5). Bad weather may make observation impossible. The animals may prove to be much more shy than had been expected and require months or years of habituation before they will allow you near enough to make useful observations. The conditions for recording behaviour in the field are rarely ideal and high-quality data are not easy to collect. In pursuing the possibility of working on an animal in the field, you need to discover whether you must overcome practical problems such as requiring special permissions from governments or landowners. How easy will it be to get to where you want to be and to live when you get there? Paterson's (2001) book describes well the kind of things that must be considered, particularly if you are planning to work in the tropics.

The days are over when a field worker could confidently suppose that good descriptions of a species obtained from one habitat could be generalised to the same species in another set of environmental conditions. Considerable differences are often found between different populations of the same species. Consequently, the field worker may find that a great deal of effort has been put into describing a special case, and the data may only start to make sense when the same species has been studied in a number of different habitats.

The special problems of studying behaviour in a natural environment mean that the field worker has to focus especially clearly on the compromises that are inevitable in all research between what is ideal and what is practicable. The balance that is struck will depend on the species being studied and the conditions in which it lives. Choice of the right species, discussed above, is especially important in fieldwork and considerable thought must be given to this matter before embarking on a study.

Although laboratory studies of animal behaviour remove the numerous confounding variables associated with field studies, certain activities may be elicited only in captivity and are therefore not part of a species' natural behavioural repertoire. A solution to this problem is to conduct controlled field experiments, so that certain aspects of the environment can be manipulated or controlled. An example of a controlled field study comes from work on egg discrimination in free-ranging weaverbirds in the Gambia, West Africa (Lahti & Lahti, 2002). In order to assess the birds' abilities to discriminate between their own eggs and those from other parents, the nests were artificially 'parasitised' with the eggs of other weaverbirds. To control for disturbance of the nests, eggs were removed from other nests, marked and replaced.

The possibilities for experimental work in the field are considerable. However, it has to be remembered that, as always, badly designed experiments are a waste of time. Moreover, all experiments are liable to be disruptive to the animals and interfere with those qualities that field studies are best equipped to investigate, namely the natural character of behaviour. While much can be done in relatively unconstrained surroundings, much cannot, and it is then necessary to use the laboratory, where conditions can be controlled and much may be known about the histories of the individual animals.

Choosing when to observe

Making observations according to some pre-determined schedule is an important precaution against the bias that would arise if you merely recorded whenever something obvious or interesting happened. In general, the times at which the recording session starts and stops should be determined in advance, and not by what the subjects are doing at the time (unless, of course, the aim of the study is to find out what happens during or after a particular type of behaviour has occurred).

Choosing the appropriate season and time of day at which to observe is an important practical issue in any study because the nature of behaviour will change with both. Obviously, animals are not equally active throughout the 24-hour period, so the amount of activity seen will depend on the time of day at which the subjects are watched. A surprising number of people study nocturnal animals such as rats when they are least active.

The problem of diurnal variations in behaviour can be approached in one of four ways:

- By recording behaviour throughout the 24-hour period, either by continuous observation or with several observation sessions spread across each day. Clearly, this is not a practical proposition in many cases, particularly if only one observer is involved. A compromise might be to observe two or three times each day; for example, during the morning and early evening. If the results obtained at the various times of day are markedly different then they must be analysed and treated separately; if not, they can be pooled to give a daily average.

- By observing at a different time on each day such that, averaged across the entire study, each part of the day is equally represented in the final sample. This approach cannot be used if there is any likelihood that the behaviour changes systematically from day to day – as, for example, when studying young, developing animals – or when behaviour undergoes marked seasonal changes. Despite good intentions of sampling uniformly throughout the day, many investigators find that their samples are in practice unevenly distributed.

- By partially ignoring the problem and observing at the same time each day. This is the most usual approach, especially in laboratory studies. Strictly speaking, if all observations are made at the same time of day then the results should not be generalised to any other time of day. In practice, this limitation on the validity of observations should not present great problems unless the time of day strongly influences the nature of the results – particularly if the aim of the study is to make comparisons between groups of subjects from the same species. Problems can, however, arise when diurnal activity rhythms drift or when comparing the behaviour of different species whose activity rhythms differ. Observations should be made at a time of day when the behaviour of interest is most likely to be occurring.

- By ignoring the problem completely and recording at a different time each day on a haphazard basis. This approach has no obvious merits, but is sometimes a necessary evil for observers studying behaviour under difficult conditions.

Comparable arguments apply to the problem of seasonal variations in behaviour.

Table 2.2 summarises the basic questions you should ask yourself before engaging in a study. Having done so, you should consider a number of other important issues.

Table 2.2 *Questions to be asked before engaging in a new study.*

What level of analysis should be used?
Which species should be used?
Where should the study be conducted?
At what time of day should the study take place?

Effects of the observer on the subject

The observer is rarely invisible to the animals being studied and may have a significant effect on them both in laboratory and in field studies. Even animals that neither react with alarm to your presence nor attempt to escape from you may nonetheless alter their behaviour in subtle ways. A strong belief that the subjects are not affected by your presence can often turn out to be mistaken.

In field studies, you can reduce disruption by using hides or blinds to conceal yourself. If you cannot approach the hide without being noticed, an accomplice may have to come with you and, when you are hidden, walk ostentatiously away – although some animals are not taken in by such a ruse. Of course, restricting yourself to a hide, even a mobile one such as a vehicle, may mean that some of the most interesting aspects of an animal's behaviour are missed. Consequently, in many studies observers spend long periods simply accustoming the subjects to their presence, a stratagem that generally seems to work well. Nonetheless, the impression that well-habituated subjects are not affected by the observer's presence is difficult to verify and should be treated with some scepticism. This point is especially relevant to field studies of behaviour.

Your presence may introduce subtle bias even though the animals appear to be well habituated. For example, some activities (such as play or sexual behaviour) or some individuals (such as juveniles) may be more affected than others by your presence. Similarly, even though the animals may be habituated to your presence, their prey (if they have prey) may not be and, worse, their predators may find it easier to catch them.

In the laboratory, a one-way screen made of half-silvered glass, dark-tinted clear plastic or translucent material can be placed between you and the animals being observed. This technique for concealing the observer relies on the illumination being much brighter on the subject's side of the screen. Another method is to place an angled mirror above the subject and, on the assumption that it does not look up, watch from a position where direct visual contact is impossible.

Of course, even if you cannot be seen, the animal may still be able to hear or smell you. We humans, with our relatively limited olfactory abilities, are prone to under-estimate the importance of smell for other species. A third possibility, which solves this problem, is to use a concealed video camera and watch the behaviour on a remote screen. The ability of video to capture detail may not be as good as direct observation, but this method has the advantage that a permanent record can be kept, it may capture rapidly occurring events that are easily missed, and the record can be analysed many times.

In laboratory studies it is generally a good idea, where possible, to study animals in the same place as they are housed. Transferring an animal from its home environment to a strange environment in order to observe it may disrupt its behaviour considerably. If testing or observing in the home cage is impossible then it may be necessary to ensure that animals are well habituated to the test situation before data are collected.

Psychologists and sociologists have known for many decades that changes in the behaviour of their human subjects sometimes result not from the effects of any experimental manipulation, but merely from the attention paid to them by the researcher. In psychology and sociology this source of error is known as the Hawthorne Effect. A striking example in studies of drugs or medical procedures is the way in which people show significant improvements in health even though they have received an inert substance or a sham procedure – the placebo effect. Animals too may be very responsive to subtle cues unconsciously given by humans, as the famous counting horse, Clever Hans, demonstrated so clearly. Its reliance on cues provided unconsciously by its trainer was only revealed when it could no longer see the trainer.

Anthropomorphism

Humans readily interpret the behaviour of other species in terms of their own emotions and intentions. Observations of animals often leave the strong impression that the animals know what they are doing. However, subsequent analysis frequently reveals that seemingly complex and purposive behaviour can be produced by simple mechanisms that do not involve conscious awareness or intentions. For example, a woodlouse will move about briskly when it is in a dry environment, and sluggishly or not at all when in a humid environment. The animal appears to seek out damp places in a purposeful manner, but its response can be explained in terms that are no more complicated than those of an electric heater controlled by a thermostat.

Using human emotions and intentions as *explanations* for animals' actions can impede further attempts to understand the behaviour. In general, therefore, you should start by obeying the injunction of Occam's razor to explain behaviour in the simplest possible way until you have good reason to think otherwise. Nonetheless, slavish obedience to such a maxim tends to sterilise imagination and, although the possibility of anthropomorphism must be acknowledged, an over-emphasis on its dangers can constrain research. If you *never* think of an animal as though it were a human you are liable to miss much of the richness and complexity of its behaviour. If an animal is invariably thought of as a piece of machinery then some of its most interesting attributes may be overlooked. A preferable approach is to muster every possible type of mental aid when generating ideas and hypotheses, but to use the full rigour of analytical thought when testing them. Attributing human sensations, emotions and intentions to an animal so that you can do more imaginative science does not mean that, when your efforts are crowned with success, you have proved that it feels and thinks like a human. The distinction between the heuristic value of such projection and its truth value needs to be explicit.

Another fundamental point is that other species occupy different perceptual worlds from humans – that is, their sensory abilities may be radically different from our own. Many rodents communicate using ultrasonic vocalisations, some insects and birds can detect ultraviolet light, some snakes can detect prey using infrared sensors and many species have highly developed powers of olfaction. Humans occupy a perceptual world that is dominated by colour vision, but this is not true for many other species. Thus, an animal may be oblivious to a visual stimulus that seems obvious to a human or, conversely, may respond to a stimulus that a human cannot detect or does not imagine is important to the animal. A colleague reported that if a person familiar to the cotton-top tamarins in a laboratory colony wore a novel pair of shoes, the monkeys responded very differently during subsequent observations.

Ethical considerations

Everyone studying animals should think carefully about the ethical implications of causing suffering and disruption to animals in the course of scientific research. Many would argue that those who study animal behaviour, yet are insensitive to the condition and welfare of their animals, lay themselves open to the charge that their science is questionable. This is not simply a matter that arises in laboratory work. Field studies of animal behaviour may also raise ethical issues if some form of experimental manipulation is involved or if the lives of the animals are disrupted.

You have a duty to abide by the spirit (as well as the letter) of any legislation governing animal research, and should strive to minimise the number of animals used in research and the amount of suffering caused to each animal. The number of subjects needed to give a clear result can sometimes be reduced by more careful choice of research design, measurement techniques and statistical analysis (see Chapter 8). In some circumstances simulated versions of animals (sometimes called animats) can be used in place of real animal subjects in experiments (Watts, 1998). The appropriate use of computer simulations in a research program can reduce the number of animals required for an experiment and also save both research time and money (see Chapter 11).

The ethical issues of using animals in research are complex and require at least three difficult types of judgement to be made (Bateson, 2005a). First, how worthwhile is a proposed piece of research in terms of its likely contributions to scientific understanding? Second, how likely is the research to bring benefits – for example to human and veterinary medicine, the economy or the environment? Finally, how much suffering to animals is likely to result from the research?

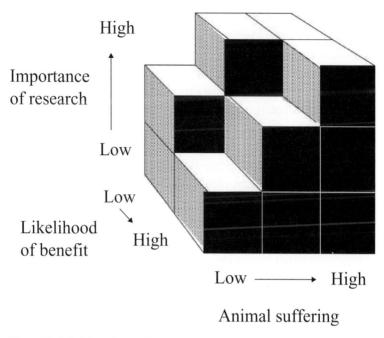

Figure 2.1 A decision cube showing how the importance of the research, the probability of benefit and the extent of the animal suffering produced in the research may be brought together. The solid part of the cube represents work that would be deemed unacceptable.

These points are brought together in the 'decision cube' shown in Fig. 2.1. Limited animal suffering is treated as acceptable only when both the importance of the research and the probability of medical benefit are assessed as being high. Moreover, certain levels of animal suffering would generally be unacceptable regardless of the quality of the research or its probable benefit. Research of high importance involving little or no animal suffering would be approved even if the work had no obvious potential benefit to humans. This recognises the need to understand phenomena that have no immediate and obvious benefit for humans. Such understanding is seen as a 'good' in itself. It is often impossible to predict how understanding biology will help to advance medicine in the future. For instance, ornithologists studying bird migration played an important role in understanding the transmission of avian flu.

While judgements about the scientific value of research may only be shared by a group with a particular interest, such assessments are regularly made by the many committees dispensing limited funds to finance scientific research and by editorial boards of scientific journals. Few would claim that such judgements are always right, but a consensus is usually reached when assigning research proposals to one of a few, broad classes of merit. A similar process is applied to judging the likely medical and other benefits.

The results of behavioural investigations of animals have had applications in a number of areas such as improving conditions of pets, farm animals and animals in zoos. They have helped in wildlife management and preservation of the environment. Studies of behavioural imprinting and other early-experience effects have led to increased understanding of the complexity of genetic and environmental interactions that produce behaviour in a variety of species, including humans. In addition, the television programmes about animal behaviour that members of the public enjoy so much are often based on the results of years of research by dedicated behavioural biologists.

Independently of judgements about scientific merit and medical benefit, a careful assessment must be made of the amount of animal suffering that is likely to be caused at all stages of the research. The subjective experiences of an animal, if it has any, may be totally different from humans, reflecting its different way of life and the different ways in which its body works. This means that human interpretation of what is observed in another animal should not be based only on extrapolations from humans but also on a good knowledge of its natural history and behaviour. Different species of animals react differently to potentially damaging situations. Stimuli that make a human run and scream might make other animals immobile. They do not look as though they are in a state of fright, because alarmed humans would not normally behave like this. With knowledge of how animals behave, the range of animals that are believed to suffer is often broadened. The plausibility of projections from human experience to other animals clearly depends on good observational data about their normal behaviour, their requirements, their vulnerability to damage and the ecological conditions in which they live.

In order to tackle some of these difficult ethical problems, the Association for the Study of Animal Behaviour (ASAB) in the UK and the Animal Behavior Society (ABS) in America formed, respectively, an Ethical Committee and an Animal Care Committee. Their joint guidelines for the treatment of animals in research and teaching are published in the January issue of *Animal Behaviour* each year.

Quite different considerations relate to observational work on humans, for whom informed consent is crucial. Are the subjects of the observational work aware of what you are doing and why you are doing it? Will their privacy be respected? In many countries the interests of human subjects are carefully protected and you may be required to submit your research proposals to an ethics committee.

Pressures to fund scientific research from private sources have increased in recent years. Many institutions will give advice on what to look out for when negotiating for such funds. It is as well to check in advance whether any restrictions will be placed on the publication of your findings and ensure that, come what may, you have a right to publish within a specified time after the completion of the work. Sometimes serious conflicts of interest may arise if you produce results that are incompatible with the interests of the funding body. Under these circumstances, you may be placed in a dilemma. Even so, you should aim to maintain your integrity as a scientist, which matters most in the long run, and think carefully about whether you have taken into account public interest (as distinct from interest by the public) when the time comes to publish.

These issues are discussed further in Chapter 11 and a checklist of questions that you should ask yourself are given in Appendix 4. Ethical issues to do with the sharing (or not) of ideas, intellectual copyright and publication are discussed at length by Hailman and Strier (2006). They will also be considered further in Chapter 11.

Summary

In planning research you should think carefully about the level at which you want to work so that you measure appropriate detail but not too much of it. You must take into account whether or not you have chosen the right subjects, where you will work on them and when. You must consider how much your presence will affect them and be careful about how much you project human intentions and emotions on to them. You must consider their particular perceptual abilities and you should be very sensitive to the ethical issues raised by working on animals and the different issues raised by working on humans. You should also consider any pressures arising from the way in which you are funded.

3

Getting started

The steps involved in studying behaviour

Recipes for conducting research are rarely followed precisely and most scientists build their ways of investigation in periods of apprenticeship when they model themselves on the behaviour of more experienced colleagues. In considering the steps listed below you should be aware that many programmes of research enter the sequence at different points. In general, though, studying behaviour involves a number of inter-related processes in roughly the sequence in which we have listed them. We have described these steps in outline. Lehner (1996) and Hailman and Strier (2006) provide much fuller accounts of research methodology, although we depart from their schemes – particularly in the emphasis we have placed on our first five steps.

1. Ask a question

Before any scientific problem is investigated, some sort of question will have been formulated. The question may initially be a broad one, stemming from simple curiosity about a species or a general class of behaviour, such as 'What does this animal do?' Such a question is not a hypothesis.

The value of broad description arising from sheer curiosity should not be under-estimated. Alternatively, it may be possible at an early stage to formulate a much more specific question based on existing knowledge and theory, such as 'Do big males of this species acquire more mates than small males?' This is tacitly a hypothesis. Not surprisingly, research questions tend to become more specific as more is discovered about a particular issue.

The particular choice of question (or questions) may be influenced by a variety of factors, including previous knowledge, interests and observations made in the course of other research and the priorities of the group in which you work. Sometimes the impetus for a study stems from little more than a hunch or from a wish to see what an animal will do next.

2. Make preliminary observations

A period of preliminary observation is generally invaluable in deciding what measurements to make and should be regarded as a crucial part of any study. Jumping straight in and collecting 'hard data' from the very beginning is rarely the best way to proceed.

3. Identify the behavioural variables that need to be measured

The form of the research and the variables that are to be measured should then be chosen so as to provide the best account of what you have observed. Definitions of behavioural categories should be clear, comprehensive and unambiguous. Write down the definitions before starting to collect data.

4. Choose suitable recording methods for measuring these behavioural variables

No observer can record behaviour without selecting some features from the stream of events and ignoring others. This selection inevitably reflects the questions you asked at the beginning of the study. It simply is not possible to record everything that happens, because any stream of behaviour could, in principle, be described in an enormous number of different ways. Practise the recording methods, assessing the reliability and validity of each category. Drop categories that are clearly unreliable and irrelevant. Measure inter- and intra-observer reliability at the beginning and end of data collection (see Chapter 7). Be prepared to add new categories and to redefine categories in the light of preliminary observations and pilot measurements.

5. Collect and analyse the data

Use the same measurement procedures throughout. Attempt to plan in advance how much data you will need to collect in order to obtain a clear conclusion (see Chapter 8). Once embarked on the collection of data, some people find it difficult to stop. Use the appropriate statistical tools for analysing the data. In some studies, this may be the point when you present your findings to a wider audience. You may also want to move to a more experimental phase, in which case the next step is crucial.

6. Formulate precise hypotheses

A clear hypothesis invites a direct test, but remember that hypotheses may be tested by observing natural variation in a population as well as by performing experiments. The study may have started with a particular hypothesis or it might have arisen from a more open-ended phase of the work covered by steps 1 to 5 above.

Formulating hypotheses is a *creative* process, requiring imagination as well as some knowledge of the issues involved. Others may already have made suggestions to explain the phenomenon of interest. It is then often a good idea to think about the weaknesses of their proposals as a guide to formulating a new hypothesis. More generally, do not reinvent the wheel: find out what others have done and speculated upon in the area and build on their work, without being so restricted by what they claim that you neglect your own ideas.

It is not possible to give definitive advice on how to formulate good hypotheses, any more than advice can be given on how to write good literature or paint good pictures. Sometimes, though, it is worth considering hypotheses that have been particularly successful in adjacent areas of research to see if they can be adapted to provide a good explanation in the area under investigation. The aim should be to find the best explanations. That means not just any old explanation compatible with the data obtained so far, but one that unifies with simplicity, giving a common account of superficially diverse phenomena.

In general, the larger the number of plausible competing hypotheses that are formulated the better, particularly when they make different predictions. The danger with having only one hypothesis is that it may be more difficult to abandon it when its predictions are not supported by the evidence. Data may always be explained in more than one way. That said, it is often a good plan to ask yourself which of the competing hypotheses would provide the most unifying explanation for the data.

7. Make predictions from the hypotheses

Making the transition from thinking about a problem and formulating hypotheses to tackling it empirically is often one of the most difficult parts of research. A clear hypothesis should, by a process of straightforward reasoning, give rise to one or more specific predictions that can be tested empirically. The more specific the predictions are, the easier it usually is to distinguish empirically between competing hypotheses, and thereby to reduce the number of different ways in which the results could be explained. A failed prediction may not necessarily mean that the hypothesis was wrong; the method of testing it may have been at fault.

8. Design the tests

Even if steps 2 to 4 have already have been passed through, it may be necessary to consider them again. The variables that are to be measured should then be chosen so as to provide the best test of the different predictions made by competing hypotheses. Experiments can be greatly improved by careful design and thoughtful use of control groups. Good design allows fewer subjects to be used overall when several treatments are combined in a single experiment (see Chapter 8). Moreover, if the treatments combine to influence the outcome, such interactions between them will be uncovered. The numbers of subjects used must, however, not be so small that the study will reveal no clear conclusions. Advice from colleagues expert in the design of experiments is almost always helpful.

9. Run tests of your hypotheses

Use the same measurement procedures throughout and try, if possible, to collect data 'blind' so that you do not unconsciously select data that fit your hypotheses. Stop collecting data when you have reached a pre-determined threshold that enables you to provide clear and reliable answers to your questions. Some inexperienced observers will stop when they get a statistically significant result, forgetting that with small sample sizes flukes are more likely to arise. On the other hand, do not go on collecting data simply because it is possible to do so. Once collected, make sure that your data are properly labelled, dated and include crucial information about the conditions and by whom the data were collected.

10. Analyse the results of your tests

Prepare the data in spreadsheets so that they are easily inspected and made available for subsequent statistical analysis. You may need to combine measures and guard against some common mistakes when treating data points as though they were independent. Employ the appropriate statistical tools, both for presenting and exploring the data, and for testing the hypotheses. Carry out *exploratory* data analysis to obtain the maximum amount of information from the data and to discover unexpected results that generate new questions. Do not sacrifice clarity for complexity by using unnecessarily complicated statistics. Use *confirmatory* analysis to test hypotheses. Distinguish between testing existing hypotheses and generating new ones (see Chapter 9).

11. Consider alternative interpretations of the evidence

Do not draw more conclusions than the data support, but do try to formulate a list of questions and ideas suggested by the data that could form the basis of future research. Be prepared to consider a range of alternatives. Try to come up with the best explanation that would, if correct, fit with background knowledge. The aim is to unify and offer coherence where it might have been lacking before.

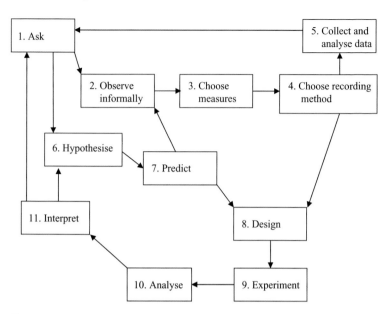

Figure 3.1 Summary of the steps involved in research. Some research will start at step 6 but would need to incorporate steps 2–4 before step 8. The progressive character of research may involve returning from step 11 to step 6 but in some cases may involve returning to step 1.

At this stage you may be ready to present your work to a wider audience orally or in written form. Always maintain scrupulous honesty. Submit your written work to peer-reviewed journals. When preparing your work for publication, pay attention to aspects of your work that may bear on matters of public interest.

As in any other area of science, measuring behaviour in order to discriminate between one set of hypotheses will inevitably produce results that in turn generate new hypotheses. In this sense, scientific research has a cyclical nature. It may be especially helpful to develop explanatory models that simplify complex phenomena, help you to understand them and to plan new work.

The steps involved in studying behaviour are summarised in Fig. 3.1. The successful scientist is likely to be one who can combine a purposeful approach to tackling an initial set of questions with the ability to recognise and respond imaginatively to new questions that arise during the course of research. A study is unlikely to be fruitful if it remains open-ended and never focuses on specific issues. Conversely, if one problem is pursued in a rigid and inflexible way to the exclusion of all else then potentially important new ideas and observations may be missed.

Preliminary observation

Quantitative recording of behaviour should be preceded by a period of informal observation, aimed at understanding and describing both the subjects and the behaviour you intend to measure. Preliminary observation is important for two reasons: first, because it provides the raw material for formulating questions and hypotheses; and second, because choosing the right measures and recording methods requires familiarity with the subjects and their behaviour. Preliminary observation is especially important if the problems or animals are new to you.

Even in relatively open-ended studies, what you observe will reflect, in part, existing knowledge and theories, as well as your own preconceptions. Measurement of behaviour depends greatly on your being familiar with the animals you will observe, both from direct experience of watching them and by reading the literature about their biology and behaviour. A period of preliminary observation also provides a valuable opportunity for sharpening up your ideas and practising recording methods.

We cannot over-state the importance of simply watching before starting to measure systematically. This is particularly important when you are working with a species that has not been studied much previously. Beginners may falter early in a study because they rush to obtain 'hard data' and do not allow sufficient time to watch, think and frame interesting questions. Even an experienced observer needs to spend time on preliminary observation.

As a practical guideline, we suggest that you should always *plan* to exclude data obtained during the first few recording sessions from your final analysis. Otherwise it can be tempting to use all the data that you obtained, even though data from early recording sessions may be unreliable or not comparable to later data because of 'observer drift' or deliberate changes in measurement procedures. After a period of trial recording sessions, in which behavioural categories and measurement techniques have been tried out, preliminary data should be analysed. It is at this stage that methods should be modified if necessary.

Describing behaviour

Behaviour can be described in a number of different ways. The simplest distinction is between describing behaviour in terms of its structure or consequences.

The **structure** is the appearance, physical form or temporal patterning of the behaviour. The behaviour is described in terms of the subject's posture and movements.

The **consequences** are the effects of the subject's behaviour on the environment, on other individuals, or on itself. In this case, behaviour may be described without reference to how the effects are achieved. Categories such as 'obtain food' or 'escape from predator' are described in terms of their consequences, and may be scored irrespective of the actual pattern of body movements used.

For example, 'turn on light' is a description in terms of consequences, while 'press switch down using index finger' is a structural description. Similarly, 'run tip of bill along primary feather of wing' is a structural description, while 'preen' is a description by consequence.

Describing behaviour by its structure can sometimes generate un-necessary detail and place demands on your ability to make subtle discrim-inations between complex patterns of movement. Description by conse-quence is often a more powerful and economical approach, and has the additional advantage that the consequences can sometimes be recorded using automatic devices.

It is not uncommon for behaviour to be described in terms of presumed consequences or causes that later turn out to be wrong. Because of this danger it is best to use neutral terms for labelling behaviour patterns, rather than labels that falsely imply knowledge of the animal's internal state or the biological function of the behaviour pattern. For example, if a category of vocalisation is named *distress call* (rather than given a neutral label such as *peep*), then you might be tempted to include vocalisations that did not meet the stated criteria for the category, but which were emitted when the animal was apparently distressed.

A third form of description is in terms of the individual's **spatial relation** to features of the environment or to other individuals. In this case, the subject's position or orientation relative to something (or someone) is the salient feature. In other words, the emphasis is not on what the subject is doing, but where or with whom. For example, 'approach' or 'leave' might be defined in terms of changes in the spatial relation between two individuals.

Choosing categories

Behaviour consists of a continuous stream of movements and events. Before it can be measured, this stream must be divided up into discrete units or categories. In some cases behaviour appears to be composed of natural units of clearly distinguishable, relatively stereotyped behaviour patterns such as pecks or grunts, and the division process will partly be dictated by the behaviour itself. To a large extent, though, the suitability of a category depends on the question being asked rather than on some inherent feature of the behaviour. Observational categories must reflect some sort of implicit theory and do not have an existence of their own, independent of the observer. It is therefore difficult to give specific advice on what sorts of categories to choose, although we can offer some general guidelines.

- Enough categories should be used to describe the behaviour in sufficient detail to answer the questions and, preferably, to provide some additional background information.
- Each category should be precisely defined and should summarise as much relevant information as possible about the behaviour.
- Categories should generally be independent of one another; that is, two or more categories should not be merely different ways of measuring the same thing.
- Categories should generally be homogeneous: that is, all acts included within the same category should share the same properties.

Inexperienced observers often err on the side of trying to record too much. A given stream of behaviour could potentially be described in an almost limitless number of ways, depending on the questions being asked, so it is essential to be selective. It is certainly best to drop categories that are clearly irrelevant, or which seem inconsistent and difficult to measure reliably. The chances are that the fewer categories used, the more reliably each will be measured.

Bear in mind, though, that you will improve with experience, so data from later recording sessions may be reliable even if data from early sessions are not. Furthermore, it may be better to record too much initially, rather than too little. Redundant or unreliable categories can always be discarded or pooled at the analysis stage. It is also wise to collect supplementary information that might in the future provide useful background or raise new questions. However, collecting a wide range of measures or supplementary information should not be allowed to detract from careful measurement of the important things.

The extent to which the definitions of individual categories are specific rather than general will depend on the nature of the problem. Questions and hypotheses tend initially to be rather broad and then narrow down as more is discovered about a particular problem. The more clearly and precisely the initial question has been formulated, the more obvious it will be what to measure.

When choosing categories it can sometimes be helpful to have descriptions of the main types of behaviour pattern that typify the species. In some cases this information is available in the form of an 'ethogram', which is ostensibly a catalogue of descriptions of the discrete, species-typical behaviour patterns that form the basic behavioural repertoire of the species. Unfortunately, published ethograms vary enormously in the number of behavioural categories included and the detail with which these are described, and ethograms are unavailable for many commonly studied laboratory subjects. Moreover, ethograms are frequently of limited use because not all members of a species behave in the same 'species-typical' way. On the contrary, individuals of the same species, even when of the same sex and age, can behave in quite different ways.

Defining categories

Each category of behaviour to be measured should be clearly, comprehensively and unambiguously defined, using criteria that can be easily understood by other observers. More important still, the criteria used to define a category should unambiguously distinguish it from other categories, particularly those it resembles most closely. A detailed and complete definition of each category and the associated recording method should be written down *before* the data used in the final analysis are collected. Two types of definitions are used:

Operational definitions specify the physical operations that are required by the researcher to make the measurement; these are most commonly used when measuring the consequences of behaviour.

Ostensive definitions involve giving an example of the case to which the category applies using diagrams and written descriptions, for example when an individual plays with an object. These definitions are most often used in the direct observation of behaviour. An ostensive definition of an action should enable another observer to recognise the same pattern of behaviour.

The period of preliminary observation provides an opportunity to develop the precise criteria used to define each category. A completely satisfactory and unambiguous definition of a category can rarely be formulated without having watched the behaviour for some time. Preliminary definitions are often unable to deal with unforeseen ambiguous examples of the behaviour that crop up during preliminary observations, and must be therefore modified in the light of experience.

Clearly, all the data for a particular category that are used in the final analysis must be strictly comparable. Thus, data obtained before the final definition of a category was formulated must be discarded. Developing a set of precise and unambiguous category definitions can be a slow process.

Writing down precise definitions of categories at the beginning of the study is essential to prevent definitions and criteria from 'drifting' during the course of the study (see Chapter 7 on factors affecting reliability). Written definitions should be sufficiently precise and detailed to enable another observer to record the same things in the same way.

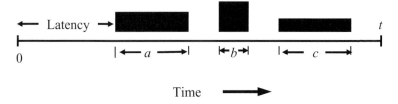

Figure 3.2 The meaning of latency, frequency, duration and intensity. The black rectangles represent three successive occurrences of a behaviour pattern during an observation period of length t units of time. Latency is the time from the beginning of the observation to the first occurrence of the behaviour. Frequency is the total number of occurrences divided by the total observation time ($3/t$). The total duration of the behaviour is $a + b + c$ units of time and the mean duration is the total duration divided by three. Intensity is the amplitude of the behaviour represented by the height of the rectangles.

Types of measure

Behavioural observations most commonly yield four basic types of measure (see Fig. 3.2).

Latency (measured in units of time: e.g. s, min or h) is *the time from some specified event* (for example, the beginning of the recording session or the presentation of a stimulus) *to the onset of the first occurrence of the behaviour*. For example, if a bird does not approach a novel object until 6 min have elapsed, the latency to approach is 6 min. If, as is normally the case, the period of observation is limited and each individual is tested more than once, then the behaviour pattern may not occur at all during some tests. A related measure to latency is the lag between one event and another – say between one animal performing an act and another performing the same act.

Frequency (measured in reciprocal units of time; e.g. s^{-1}, min^{-1} or h^{-1}) is *the number of occurrences of the behaviour pattern per unit time*. Frequency is a measure of the *rate* of occurrence. For example, if a rat presses a lever 60 times during a 30-min recording session, the frequency of lever pressing is 2 min^{-1}.

An alternative usage, which is perhaps more common in the behavioural literature and in statistics, is when 'frequency' refers to the *total number of occurrences*. However, this usage is uninformative and potentially misleading, unless the total time for which the behaviour was watched is also specified. For example, to state that the 'frequency' of a behaviour was 60 is meaningless: did it happen 60 times in two minutes; one hour; a day . . .? Most statements about total numbers of occurrences could equally well refer to rates of occurrences, since a total number of occurrences can always be expressed as a rate, assuming the length of the observation period is known. To avoid any confusion, the *total number of occurrences* should be explicitly referred to as such. Expressing frequencies in the way we suggest (number per unit time) removes any possible ambiguity.

Duration (measured in units of time: e.g. s, min or h) is *the length of time for which a single occurrence of the behaviour pattern lasts.* For example, if a kitten starts suckling and stops 5 min later, the duration of that period of suckling was 5 min.

'Duration' is also used in at least two other senses in the behavioural literature. The first is when 'duration' (or 'total duration') refers to the *total* length of time for which all occurrences of the behaviour lasted over some specified period, usually the whole observation session. A total duration is, of course, meaningless unless the total time for which the behaviour was watched is also specified. For example, to state that the 'total duration' of a behaviour pattern was 16 min says nothing: was it 16 min out of 20 min; 30 min; an hour; a day . . .? To avoid any ambiguity, we recommend that a total duration should be expressed as the total duration over the specified period of observation (for example, '9 min per 30 min') and should be explicitly referred to as **total duration**.

Alternatively, a total duration can be expressed as a proportion (or percentage) of the observation period, in which case it should be explicitly referred to as the **proportion** (or percentage) of time spent performing the behaviour. For example, if a kitten spent a total of 10 min suckling during a 30-min observation session, then the proportion of time spent suckling was $10/30 = 0.33$. Note that a proportion or percentage of time is a dimensionless index with no units of measurement.

Expressing a duration as a proportion or percentage of total time omits the potentially important information about the total time for which the behaviour was watched. For example, the interpretation placed on the statement that the proportion of time spent sleeping by a subject was 0.10 must depend on whether this figure was based on, say, a 24-h period of observation as opposed to a 30-min observation.

'Duration' (or 'mean duration') is also used to refer to the *mean* length of a single occurrence of the behaviour pattern, measured in units of time (e.g. s, min or h). This is obtained by recording the duration of each occurrence of the behaviour pattern and calculating the mean of these durations. To avoid any possible ambiguity, we suggest that this measure should be referred to explicitly as a **mean duration**.

Mean duration can also be calculated by dividing the total duration of the behaviour pattern by the total number of occurrences. This has the advantage that the duration of each occurrence need not be recorded separately; you could, for example, use a cumulative stopwatch to record the total duration and a counter to record the total number of occurrences.

As an illustration of these various measures, suppose that a mammalian mother and offspring are observed for 60 min, during which suckling occurred five times, the individual periods of suckling lasting 3 min, 10 min, 1 min, 1 min and 1 min, respectively. According to our suggested definitions, the *durations* of suckling were 3,10,1,1 and 1 min; the *total duration* of suckling was 16 min per 60 min; the *proportion of time* spent suckling was 0.27 (= 16/60); and the *mean duration* of suckling was 3.2 min (= 16/5).

Frequency and duration, which are the measures most commonly used for describing behaviour, can give different and complementary pictures. For example, how often two monkeys groom each other (frequency) tells us something different about the nature of their social relationship from how long they spend doing it (duration). Frequency and duration measures of the same behaviour are not always highly correlated, so it is probably wise to record both.

Intensity. In general, categories are best defined in such a way that the behaviour is simply scored according to whether or not it has occurred or for how long it has occurred, rather than making assessments of intensity or amplitude. Intensity, unlike latency, frequency and duration, has no universal definition. Nonetheless, it may be helpful or even essential to make judgements about the intensity or amplitude of a behaviour pattern.

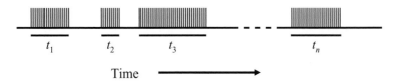

t_1 t_2 t_3 t_n

Time

Figure 3.3 The meaning of local rate, which provides another measure of the intensity of behaviour. An activity such as eating may be composed of discrete component acts such as ingesting a morsel of food, indicated here by vertical lines. The local rate is given by the total number of occurrences of the component act (number of food morsels eaten during the observation period) divided by the total duration of the activity ($t_1 + t_2 + t_3 + \ldots + t_n$)

In some cases the consequences of the behaviour can be measured in terms of some physical quantity related to the behaviour, such as the weight of food eaten, the volume of water drunk, the number of prey captured or the distance travelled. It may be useful to measure the sound intensity of a vocalisation, the amplitude of a limb movement or the height of a jump; or to estimate the intensity of a facial expression or the aggressiveness of a social interaction. Intensity can sometimes be measured according to the presence or absence of certain components of the act, which may be present at high intensity but absent at low intensity.

A simple and informative index of intensity is **local rate**, defined as the number of component acts per unit time spent performing the activity. For example, suppose that an activity – eating – is composed of discrete, component acts – the ingestion of individual items of food. The local rate of eating would, in this case, be given by the number of items ingested per unit time spent eating (see Fig. 3.3). Similarly, the intensity of walking might be measured by the number of strides per unit time spent walking, and the intensity of grooming by the number of face-stroking paw movements per unit time spent grooming. Local rate captures the speeded-up or hurried nature of intense behaviour: the more hurriedly the activity is performed, the higher its local rate.

Events and states

When choosing the type of measure to describe a behaviour pattern, it is helpful to distinguish between two fundamental types of behaviour pattern which lie at opposite ends of a continuum.

Events are behaviour patterns of relatively short duration, such as discrete body movements or vocalisations, which can be approximated as
points in time. The salient feature of events is their *frequency* of occurrence. For example, the number of times a dog barks in one minute would
be a measure of the frequency of a behavioural event.

States are behaviour patterns of relatively long duration, such as prolonged activities, body postures or proximity. The salient feature of states
is their *duration* (mean or total duration, or the proportion of time spent
performing the activity). For example, the total time a dog spends asleep
over a 24-h period would be a measure of the total duration of a state.
(Note that the term 'state' is also used in the behavioural literature to refer
to a motivational state, such as hunger or thirst, so it is important not to
confuse the two.)

The onset or termination of a behavioural state can itself be scored as
an event and measured in terms of its frequency.

The different levels of measurement

Measurement means assigning numbers to observations according to
specified rules. Four different levels of measurement are distinguished,
in ascending order of strength of measurement:

Nominal. If observations are simply assigned to mutually exclusive,
qualitative classes or categories, such as male/female or active sleep/quiet
sleep/awake, then the variable is measured on a nominal (or categorical)
scale. If only two outcomes are possible (e.g. yes/no, or male/female) the
data are said to be binary.

Ordinal. If the observations can also be arranged along a scale according to some common property then the variable is measured on an ordinal
(or ranking) scale. The number assigned to each measurement is its **rank**.
For example, if in a given period of time individual *A* played more than *B*
who played more than *C*, then *A* would be ranked highest on this measure,
B next and *C* lowest.

Interval. If, in addition, scores can be placed on a scale such that the
distance between two points on the scale is meaningful – i.e. the *difference*
between two scores can be quantified – the variable is measured on an

interval scale. The zero point and unit of measurement are arbitrary for an interval scale. A temperature measured in degrees Celsius is measured on an interval scale.

Ratio. The highest level of measurement is attained when the scale has all the properties of an interval scale but also has a true zero point. This is referred to as a **ratio** scale since, unlike an interval scale, the ratio of any two measurements is independent of the unit of measurement. Mass, length and time are measured on ratio scales. For example, the ratio of two periods of time is the same whether they are measured in seconds or days. True frequencies, durations and latencies are measured on ratio scales.

Assigning numbers to a category of behaviour does not necessarily mean that the behaviour is measured on an interval or ratio scale. For example, subjectively rating an individual's aggressiveness on a scale of 0 to 5 would not constitute measurement on a true interval scale, since there is no reason to assume that the difference between scores of, say, 1 and 2 is the same as the difference between scores of 4 and 5. Such scores can be ranked, but the differences between them are probably not meaningful and therefore the measurement would be on an ordinal scale. A ratio scale is simply not attainable in much behavioural work.

Summary

The sequence of operations required in most studies of behaviour is shown in Fig. 3.1. Attempts to test a specific hypothesis need not be involved at the outset of a study. Later in the sequence, however, hypotheses will be formulated and these should lead to direct tests. A period of preliminary observation is essential in any study. How the activities of interest are described and categorised will depend on the questions that are being asked. So too does the level of measurement.

4

Individuals and groups

Identifying individuals

In many studies, being able to identify individuals is essential. Focusing on the behaviour of an individual in a group is virtually impossible without a way of distinguishing reliably between one individual and another. Moreover, when differences in behaviour of known individuals are recorded, the resulting data are likely to be much more informative. Only by identifying and watching individuals does it become clear that all individuals in a species do not behave in the same 'species-typical' way.

In the laboratory, identification of individuals by rings, tags, collars, tattoo marks, painting the skin, dying feathers, fur-clipping and so forth does not usually offer major practical difficulties. However, it is important to realise that marking an individual may alter its behaviour or that of other individuals with which it interacts. To give one example, research revealed that coloured plastic leg bands placed on male zebra finches affected how attractive they were to members of the opposite sex. Female zebra finches preferred males wearing red leg bands over unbanded males, while males preferred females with black leg bands. Both males and females tended to avoid members of the opposite sex wearing green or blue leg bands (Burley, 2006). These findings clearly show that for zebra finches, and probably many other species, methods conventionally used to identify them can have a significant effect on behaviour.

In field studies, trapping and marking can present formidable problems. Capturing animals using traps, nets or stupefying drugs can be difficult and some forms of marking do not last long under field conditions. If traps are not monitored regularly for captured animals, any captures can fall victim to various hazards, including dehydration, starvation, exposure to adverse climatic conditions and predation. Careful thought should be given to minimising the distress caused by marking animals, for scientific as well as ethical reasons. Shooting a tranquilising dart into an animal to capture it carries many risks, both for the animal and for the humans involved in the darting. Darting can also adversely affect the social behaviour of the group under study, thereby invalidating subsequent observations for some time after darting has taken place. Therefore anaesthetic darting should only be undertaken when the reasons for doing so are strong; top priority should be given to the safety and care of animals and people. For a review of the darting of free-ranging animals see Sapolsky and Share (1998).

Various techniques enable animals to be tracked over large distances using miniature radio transmitters or sources of low-level radioactivity attached to the subject (Kenward, 2000). For most mammals and birds a tolerable weight limit for radio-collars should be under 5% of the body weight. Miniature radio-telemetry systems can discriminate automatically between individuals and are suitable with small animals, both in the laboratory and in the field. With the development of longer lasting and smaller power sources, a wider range of species can be radio-tracked in a greater diversity of habitats. Collars can be fitted with GPS (Global Positioning System) and small data-loggers. The data-loggers may be released remotely from the collars and subsequently recovered. The technology is improving all the time.

In some species, individuals have naturally distinctive markings. For example, zebras' stripes are like human fingerprints, no two individuals being identical. Similarly, gorillas' noses, elephants' ears, the whisker spots of lions, the fins of dolphins, cheetahs' tails and the bills of Bewick's swans, to list only a few, are highly variable, enabling experienced observers to recognise individuals. Many animals living in the wild acquire distinctive marks through injury, such as torn ears, damaged tails, scars or stiff limbs. Again, features such as these can be used to distinguish one individual from another. Identifying animals by naturally occurring features can be difficult and requires patience and practice, but it is the best approach in terms of minimising suffering and disruption.

When experienced observers are convinced that they can recognise individuals without recourse to written records, some verifiable demonstration of their ability is advisable. One technique is to photograph the animals as they are simultaneously identified by the observer. The tester then removes from each photograph additional environmental cues that might be helpful in identification, and records the identity of each individual. Days or weeks later, the tester presents the observer with the pictures in random order and asks the observer to name each individual.

Individual differences

The statistical techniques that are normally used for analysing behavioural data are designed primarily for drawing inferences about populations rather than individuals. A common aim therefore is to iron out the troublesome effects of individual differences in behaviour and to emphasise what members of a population have in common, rather than how they differ. However, some of the behavioural differences between individuals are of considerable biological significance and cannot be dismissed as mere statistical noise. Caterpillars are faced with very different requirements from those of the adult butterflies and moths of the same species. Female mammals caring for their young have very different priorities from those male mammals that play no part in parental care. Individuals of the same age and sex may exhibit qualitatively different modes of behaviour because they differ genetically or because they or their parents have developed under different environmental conditions (Jablonka & Lamb, 2005).

Behavioural biologists have long since abandoned the notion that behavioural characteristics may be generalised to all members of a species or even those of the same age and sex. Functional and evolutionary analysis suggests that much variation in behaviour within members of the same species of the same age and sex has been maintained by Darwinian evolution. Two or more types of reproductive tactic such as holding a harem or sneaking copulations may co-exist within populations of a given species.

Many animals are equipped with developmental plasticity so that they respond to conditions of early life, such as the nutritional state of their mother in mammals, and develop in a way that is appropriate to the conditions predicted for their adulthood; this is also true of humans (Bateson *et al.*, 2004).

Given that biologically meaningful variation in behaviour occurs so frequently, a study involving the measurement of behaviour must be planned to take such variation into account. In practice, what this amounts to is categorising individuals in advance and, better still, identifying them. When that is done, the measurements can subsequently be grouped according to category or examined to see whether they fall into obviously discrete sub-populations. If they do not, the measurements may be lumped subsequently.

Assessing individual distinctiveness

Many people who spend time watching the same animals come to feel that individuals have distinctive personalities. These impressions can often be rated reliably and validly. Data derived from observers' evaluations of individuals are used routinely in studies of human personality. An individual's personality characteristics are generally regarded as those stable, internal factors that predispose him or her to behave consistently over time and in ways that distinguish that person from other individuals.

Adult human personality can be broadly captured by five factors: Extroversion, Agreeableness, Conscientiousness, Emotional stability (or Neuroticism) and Openness to experience. These measures are found to correlate with a range of other observable outcomes, including physical health, mental health, quality of personal relationships, happiness, choice of occupation, performance at work, political views and criminality (Ozer & Benet-Martinez, 2006). Research on behavioural style in children generally refers to *temperament*, defined as 'constitutionally based individual differences in behavioural style that are visible from early childhood' (Sanson *et al.*, 2002).

Observers' ratings of animals can also provide useful information about subtle aspects of an individual subject's style that is not easily obtained in other ways. Rating methods of animals' personalities can capture the overall pattern of an individual's behaviour that remains elusive when discrete events are measured. The overall pattern involves behaviour occurring in a wide variety of conditions and it may also take into account what happens in complex social interactions. For example Wemelsfelder, Hunter, Mendl *et al.* (2001) gave nine observers freedom to choose their own descriptive terms when watching the behaviour of pigs. The independent ratings showed significant agreement in the terms applied to each pig such as 'confident', 'nervous', 'calm' or 'excitable'. One of the most successful and reliable distinctions applied to personality differences in animals of a variety of different species is between individuals that are bold and those that are timid (Carere & Eens, 2005).

Defining a group

In practice, the rules for defining groups are usually implicit: groups can often be defined intuitively by assessing how the animals are distributed in space and observing the relative distances between individuals. Those below a certain distance are regarded as being within a group, and those above it as outside.

The distance used to decide whether animals are within a detectable distance from each other depends on species-specific sensory capabilities and the environment they inhabit. For example, more cryptic species – such as lemurs – that tend not to emit loud vocalisations and occupy forest environments, will have a smaller detection distance than larger more vocal species occupying open environments, such as baboons. This informal approach can cause difficulties if the groups are not tightly clustered or if individuals spend only some of their time together. It is important therefore to make the rules for defining a group as explicit as possible, and to use carefully chosen criteria when deciding on the critical distances and minimum time spent with others that define membership of a group. A distinction is sometimes drawn between **groups** – associations whose composition is known – and **parties**, which are aggregations whose membership is uncertain.

One way of deciding on the criterion distance for defining membership of a group or party is to calculate the distance to the nearest neighbour, which might be defined as the one whose head is closest to the head of the target animal. This procedure is repeated systematically for each visible individual, excluding previous target animals. If the nearest-neighbour distances are distributed bimodally, the gap between those that are close to each other and those that are further away defines the group or party. The definition of 'together' will depend on the study: it might require that the animals are in bodily contact; that they are within one body-length of each other; or even that they are merely within visual contact. We should emphasise that this is but one of many ways to define a group or party, and that the problems of definition vary hugely between species. What is appropriate for red deer, say, may not be appropriate for fish or children. In whatever way a group is defined, the collective pattern of many animals' behaviour may sometimes provide the most appropriate unit of measurement.

Summary

In many studies individuals must be identified. Such identification is particularly important when individuals differ greatly from each other, either because of genetic (or epigenetic) differences or because their early environments induced substantial differences in phenotypes. Such behavioural variation in animals may be likened to personality differences in humans. In some studies the social group must be defined.

5

Recording methods

Sampling rules

When deciding on systematic rules for recording behaviour, two levels of decision must be made. The first, which we refer to as **sampling rules**, specifies which subjects to watch and when. This covers the distinction between ad libitum sampling, focal sampling, scan sampling and behaviour sampling. The second, which we refer to as **recording rules**, specifies *how* the behaviour is recorded. This covers the distinction between continuous recording and time sampling (which, in turn, is divided into instantaneous sampling and one-zero sampling; see Fig. 5.1). Do not use 'focal (animal) sampling' as a synonym for continuous recording described below. To do so would conflate a sampling rule (which individual is watched) with a recording rule (how behaviour is recorded).

In this section, we consider the four different sampling rules.

Ad libitum sampling means that no systematic constraints are placed on what is recorded or when. You simply note down whatever is visible and seems relevant at the time.

Clearly, the problem with this method is that observations will be biased towards those behaviour patterns and individuals which happen to be most conspicuous. For example, ad libitum sampling tends to miss brief responses and underestimates the involvement of some age groups in social interactions (Hernández-Lloreda, 2006). Provided this important limitation is borne in mind, however, ad libitum sampling can be useful during preliminary observations, or for recording rare but important events.

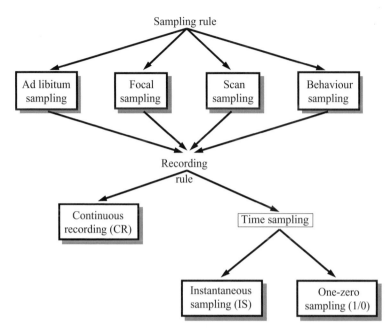

Figure 5.1 The hierarchy of sampling rules (determining which individual is watched and when) and recording rules (determining how behaviour is recorded).

Focal sampling (or focal animal sampling) means observing one individual (or one dyad, one litter or some other unit) for a specified amount of time and recording all instances of its behaviour – usually for several different categories of behaviour. Ideally, the choice of focal individual is determined prior to the observation session and the sequence in which individuals are watched should be varied systematically. When recording the social behaviour of the focal individual it may also be necessary to record certain aspects of other individuals' behaviour, such as who initiates interactions and to whom behaviour is directed. Focal sampling is generally the most satisfactory approach to studying groups.

Sometimes the focal individual will be partially obscured or move completely out of sight, in which case recording has to stop until it is visible again. Any such interruption should be recorded as 'time out', and the final measure calculated according to the time for which the focal individual was visible. Note, however, that omitting the time when the subject was not in view may introduce bias if subjects systematically tend to do certain things while out of sight. For example, many animals seek privacy when eating or mating, so their visible behaviour is not necessarily representative of their behaviour as a whole.

Focal sampling can be particularly difficult under field conditions because the focal individual may leave the area and disappear completely. If this happens, should it be pursued and, if so, for how long; or should the observations be stopped and a new focal individual chosen? Explicit rules are needed for deciding what to do if the focal individual does disappear in the middle of a recording session.

Scan sampling means that a whole group of subjects is rapidly scanned, or 'censused', at regular intervals and the behaviour of each individual at that instant is recorded. Scan sampling usually restricts the observer to recording only one or a few simple categories of behaviour, such as whether or not a particular activity is occurring or which individuals are asleep.

The time for which each individual is watched in a scan sample should in theory be negligible; in practice, it may at best be short and roughly constant. A single scan may take anything from a few seconds to several minutes, depending on the size of the group and the amount of information recorded for each individual.

Scan sampling results may be biased because some individuals or some behaviour patterns are more conspicuous than others. For some purposes, though, scan sampling has distinct practical advantages. For example, de Ruiter (1986) used it to study the activity of two groups of wild capuchin monkeys. The technique enabled him to obtain data that were evenly representative across all individuals, time of day and season, allowing various behavioural and ecological comparisons to be drawn between the two groups. Such a broad spread of data would not have been practicable had focal animal sampling been used.

Scan sampling can be used in addition to focal sampling during the same observation session. The behaviour of a focal individual may be recorded in detail, but at fixed intervals (say, every 10 or 20 min) the whole group is scan-sampled for a single category, such as the predominant activity or proximity to another individual.

If scan samples are to be used as separate data points, rather than averaged to provide a single score, they must be statistically independent of one another. This means that they must be adequately spaced out over time: clearly, scan samples taken at, say, 30-s intervals would not constitute independent measurements.

Behaviour sampling means watching the whole group and recording each occurrence of a particular type of behaviour, together with details of which individuals were involved. Behaviour sampling is mainly used for recording rare but significant types of behaviour, such as fights or copulations, where it is important to record each occurrence. Rare behaviour patterns would tend to be missed by focal or scan sampling. Behaviour sampling is often used in conjunction with focal or scan sampling and is subject to the same source of bias as scan sampling, since conspicuous occurrences are more likely to be seen. Indeed, behaviour sampling is sometimes referred to as 'conspicuous behaviour recording'.

Recording rules

Recording rules (see Fig. 5.1) are of two basic types:

Continuous recording (or all-occurrences recording) aims to provide an exact and faithful record of the behaviour, measuring true frequencies and durations and the times at which behaviour patterns stopped and started.

Time sampling involves sampling the behaviour periodically. Less information is preserved and an exact record of the behaviour is not necessarily obtained.

Time sampling in its turn can be sub-divided into two principal types: **instantaneous sampling** and **one-zero sampling**. Time sampling is a way of condensing information, thereby making it possible to record several different categories of behaviour simultaneously. In order to do this, the observation session is divided up into successive, short periods of time called **sample intervals**. The instant of time at the end of each sample interval is referred to as a **sample point** (see Fig. 5.2). For example, a 30-min observation session might be divided up into

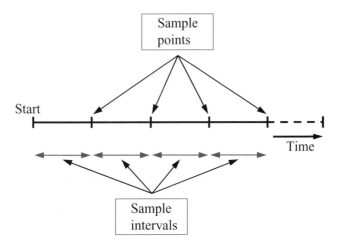

Figure 5.2　The division of an observation session into successive short units of time, or sample intervals, for purposes of time sampling. The end of each sample interval is known as a sample point.

15-s sample intervals, giving 120 sample points. Successive sample points are denoted by a stopwatch or, more conveniently, by an electronic timer that gives you an audio cue, or by a signal from an event recorder (see Chapter 6). The distinction between continuous recording and time sampling also applies when computers are used to record data automatically.

A final point worth emphasising is that continuous recording and time sampling can be used simultaneously for recording different categories of behaviour.

Continuous recording

With continuous recording (or all-occurrences recording) *each occurrence* of the behaviour pattern is recorded, together with information about its time of occurrence. True continuous recording aims to produce an exact record of the behaviour, with the times at which each instance of the behaviour pattern occurred (for events), or began and ended (for states).

For both events and states, continuous recording generally gives true frequencies, and true latencies and durations if an exact time base is used. However, bias can arise if a measurement of duration or latency is terminated before the bout of behaviour actually ends, either because the recording session ends or because the subject disappears from view. This is because the longer a bout of behaviour lasts, the more likely it is that its duration will be under-estimated by the termination of recording.

Continuous recording preserves more information about a given category of behaviour than time sampling, and should be used whenever it is necessary to measure true frequencies or durations accurately. Continuous recording is also necessary when the aim is to analyse sequences of behaviour. However, its use can be limited by practical considerations, since continuous recording is more demanding than time sampling. One consequence is that fewer categories can be recorded at any one time. Trying to record everything can mean that nothing is measured reliably.

Instantaneous sampling

Instantaneous sampling is sometimes confusingly referred to as scan sampling in the behavioural literature. It has also been called point sampling, or fixed-interval time point sampling. With this sampling method the observation session is divided into short sample intervals. On the instant of each sample point (on the 'beep'), a record is made of whether or not a given behaviour pattern is occurring (see Fig. 5.3).

The score obtained by instantaneous sampling is expressed as the proportion of all sample points on which the behaviour pattern was occurring. For example, if a 30-min recording session was divided into 15-s sample intervals, and a behaviour pattern occurred on 40 out of the 120 sample points, the score would be $40/120 = 0.33$. (Note that an instantaneous sampling score is a dimensionless index with no units of measurement.) Instantaneous sampling gives a single score for the whole recording session. Individual sample points within a session cannot be treated as statistically independent measurements.

Instantaneous sampling is used for recording behavioural states that can unequivocally be said to be occurring or not occurring at any instant

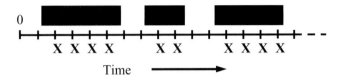

Figure 5.3 When using instantaneous sampling, occurrences of the behaviour pattern, denoted here by black rectangles, are scored with Xs at the sample points.

in time – such as measures of body posture, orientation, proximity, body contact, or general locomotor activity. Instantaneous sampling is *not* suitable for recording discrete events of short duration. Neither is it suitable for recording rare behaviour patterns, because a rare behaviour pattern is unlikely to be occurring at the instant of any one sample point and therefore will usually be missed.

One potential source of bias with instantaneous sampling is the observer's natural tendency to record conspicuous behaviour patterns even if they occur slightly before or after the sample point. The sample point is therefore stretched out from an instant to become a window of finite duration, making the sampling no longer 'instantaneous'. If, as is likely, this is mainly done with the more noticeable or important behaviour patterns then these will tend to be over-estimated relative to less prominent behaviour patterns. This type of error can sometimes be detected when two people watch the same video recording.

One-zero sampling

One-zero sampling is sometimes referred to as 'fixed-interval time span sampling'. In one-zero sampling, as with instantaneous sampling, the recording session is divided up into short sample intervals. On the instant of each sample point (on the 'beep'), you record whether or not the behaviour pattern has occurred during the preceding sample interval. This is done irrespective of how often, or for how long, the behaviour pattern has occurred during that sample interval (see Fig. 5.4). An equivalent procedure is to record the behaviour pattern when it first occurs, rather than waiting until the end of the sample interval.

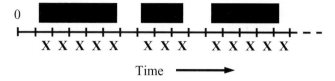

Figure 5.4 When using one-zero sampling, occurrences of the behaviour pattern, denoted here by black rectangles, within the sample intervals are scored with Xs.

The score obtained by one-zero sampling is expressed as the proportion of all sample intervals during which the behaviour pattern occurred. For example, if a behaviour pattern occurred during 50 out of the 120 15-s sample intervals in a 30-min recording session, the score would be $50/120 = 0.42$. As with instantaneous sampling, one-zero sampling gives a single, dimensionless score for the whole recording session. Again, individual sample points within a recording session cannot be treated as statistically independent measurements.

Some authorities assert that one-zero sampling should *never* be used. However, Zinner *et al.* (1997), who carefully reviewed the alternatives after using them on the same video recordings, argued that with sufficiently small sampling intervals, one-zero sampling was acceptable. Our own view is that one-zero sampling is useful for recording certain types of intermittent behaviour patterns for which neither continuous recording nor instantaneous sampling is practicable. This applies particularly to behaviour patterns, such as play in mammals and birds, which start and stop repeatedly and rapidly, and last only briefly on each occasion. Exploiting this benefit, Diamond and Bond (2004) used one-zero sampling for recording the play of two species of New Zealand parrots.

Choosing the sample interval

The size of the sample interval used in time sampling will depend on how many categories are being recorded, as well as the nature of the behaviour. The shorter the sample interval, the more accurate a time-sampled record will be. However, the shorter the sample interval, the more difficult it is to record reliably several categories of behaviour at once – especially if the activity is complicated or occurs rapidly.

In practice, you must strike a balance between the theoretical accuracy of measurement, which requires the shortest possible sample interval, and ease and reliability of measurement, which require an adequately long interval. If the sample interval is too short, observer errors can make recording less reliable than if a slightly longer interval were chosen. Thus, the sample interval should be the shortest possible interval that allows you to record reliably, after a reasonable amount of practice.

The best sample interval depends on what is being measured and is partly a matter of trial-and-error. To give some idea, many observers use a sample interval in the range from 10 s to 1 min, with sample intervals of 15, 20 or 30 s being common under laboratory conditions. Field studies, especially those in which long recording sessions are used, may require longer sample intervals.

Rather than relying on trial-and-error, the sample interval can be chosen objectively, although this requires a considerable amount of additional work. First, a fairly large sample of the behaviour must be measured using continuous recording, to give a true picture of what actually happened. Scores are then calculated for each category as though the behaviour had been recorded using time sampling with various sample intervals (e.g. 10, 20, 30 . . . s). The discrepancy between the continuous record and the simulated time sampling measure can then be calculated for each sample interval. The error caused by time sampling will increase as the simulated sample interval becomes larger. However, it may be possible to distinguish an obvious 'break-point', above which time sampling is too inaccurate, but below which it gives a reasonable approximation to continuous recording. This point marks the longest sample interval that can be used if the record is to be reasonably accurate for that category of behaviour.

In most studies, of course, several different behavioural categories are recorded and the sample interval used must be suitable for all of them. The simplest approach here is to specify the maximum acceptable discrepancy between the continuous record and the time-sampled measure for any category (say, 10%) and plot the proportion of categories where this condition is satisfied as a function of the sample interval.

One problem with this whole process is the need initially to measure the behaviour using continuous recording in order to provide a true record for comparison. In many cases time sampling is used precisely because continuous recording is simply not practicable, which would obviously rule out this procedure.

The disadvantages and advantages of time sampling

Time sampling methods are not perfect. Neither instantaneous sampling nor one-zero sampling gives accurate estimates of frequency or duration unless the sample interval is short relative to the average duration of the behaviour pattern (see Fig. 5.5) – although analytical techniques are available for deriving unbiased estimates of frequency and duration from time-sampled measures. Moreover, time sampling is not generally suitable for recording sequences of behaviour unless the sample interval is very short. This is because with one-zero sampling two or more behaviour patterns can occur within the same sample period, while instantaneous sampling can miss changes in behaviour that occur between sample points.

Instantaneous sampling does not give true frequencies or durations. However, if the sample interval is short relative to the average duration of the behaviour pattern then instantaneous sampling can produce a record that approximates to continuous recording. The shorter the sample interval, the more accurate instantaneous sampling is at estimating duration and the more closely it resembles continuous recording. If the sample interval is short then an instantaneous sampling score gives a direct estimate of the proportion of time for which the behaviour occurred. For example, if the instantaneous sampling score were 0.25, the best estimate of the proportion of time spent performing the behaviour would also be 0.25.

The accuracy of instantaneous sampling depends on the length of the sample interval (which should be as short as possible), the average duration of the behaviour pattern (which should be long relative to the

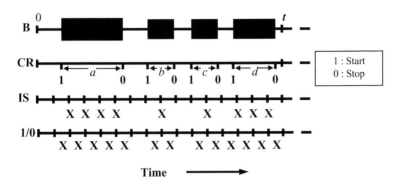

Figure 5.5 Comparison between continuous recording (CR) and the two types of time sampling: instantaneous sampling (IS) and one-zero sampling (1/0). The black rectangles on the upper trace represent four successive occurrences of a behaviour pattern (B), during an observation period lasting t units of time and divided into 16 sample intervals. Assuming arbitrarily that each sample interval lasts 10 units of time, then the following scores would be given by the three recording rules:

Continuous recording: Total duration $= a + b + c + d = 39 + 16 + 12 + 28 = 95$ units
of time. Mean duration $= 95/4 = 23.8$ units of time.
Proportion of time spent performing the behaviour pattern $= 95/160 = 0.59$. Frequency $=$
$4/t = 0.025$ per unit time. Total number of bouts $= 4$.
Instantaneous sampling: Score $= 9/16 = 0.56$
One-zero sampling: Score $= 13/16 = 0.81$

Note that instantaneous sampling gives a reasonably close approximation to the proportion of time spent performing the behaviour pattern (0.56 versus 0.59) and accurately records that four separate bouts occurred. One-zero sampling considerably over-estimates the proportion of time spent (0.81 versus 0.59) and records only three separate bouts.

sample interval) and, strictly speaking, the average duration of the interval between successive bouts of the behaviour (which should also be long relative to the sample interval). Of course, if a very short sample interval is used then the practical benefits of time sampling are lost, in which case continuous recording might as well be used instead.

Time sampling can in practice give reasonable estimates of durations or frequencies, depending on the particular characteristics of the behaviour. One-zero sampling is usually less satisfactory than instantaneous sampling and does not give true or unbiased estimates of durations or frequencies. The proportion of sample intervals in which the behaviour occurred to any extent cannot be equated either with the length of time spent performing the behaviour, or with the number of times the behaviour occurred. One-zero sampling consistently *over*-estimates duration to some extent, because the behaviour is recorded as though it occurred throughout the sample interval when it need not have done so. One-zero sampling also tends to *under*-estimate the number of bouts performed, because the behaviour could have occurred more than once during a sample interval. The shorter the sample interval relative to the average duration of the behaviour pattern, the more closely one-zero sampling approximates to instantaneous sampling.

If one-zero scores are compared – either between subjects or for different occasions – then problems can arise unless the mean bout length of the behaviour remains roughly constant. This is because the error in estimating frequency or duration depends on the ratio of mean bout length to sample interval. Thus, if the mean bout length of the behaviour varies between individuals (or, for the same individual, varies between different recording sessions) then the error in estimating frequency or duration will also vary The extent of the discrepancy between recordings depends on the relationship between the mean bout lengths and the sampling interval. For example, if a sampling interval of one unit had been used, one bout of ten units in duration would have spanned a maximum of 11 sample intervals in which the behaviour could have been recorded. In contrast, five bouts of two units in duration would have spanned a maximum of 15 sample intervals in which the behaviour could have been recorded, even though the total duration was the same as in the first recording. The discrepancy would depend on the relative durations of the mean bout intervals and their relationship to the sampling intervals. A wise precaution, in any event, is to check the mean bout lengths when you compare two recordings.

A major practical advantage of time sampling is that, by condensing the information recorded and thereby reducing your workload, it allows a larger number of categories to be measured than is possible with continuous recording. This can be an important consideration, especially in a preliminary study where it may be necessary to record a large number of categories. Time sampling also allows a larger number of subjects to be studied, if the individuals in a group are watched cyclically (i.e. using a scan sampling procedure). For instance, to record the behaviour of a group of 12 animals you might look at them cyclically, noting the behaviour of each individual (using instantaneous sampling) every 15 s, thereby watching each individual once every 3 min.

Because it is simpler and less demanding than continuous recording, time sampling also tends to be reliable. Some types of behaviour occur too rapidly for each occurrence to be recorded, making time sampling a necessity. The practical benefits of time sampling are, of course, achieved at the expense of preserving less information about the behaviour than is the case with continuous recording.

As we noted above, one-zero sampling is arguably the only practicable method for recording a type of behaviour encountered in many studies, namely behaviour patterns that start and stop repeatedly and rapidly, and last only briefly on each occasion – such as the play behaviour of some young mammals and birds. In such cases, continuous recording or instantaneous sampling are not practicable, since it is difficult (or impossible) to record each occurrence of the behaviour or to specify at any one instant whether or not the behaviour is occurring; however, it usually *is* possible to state unequivocally whether or not the behaviour has occurred during the preceding sample interval.

One-zero scores are valid measures of behaviour in their own right, in so far as they provide a meaningful index of the 'amount' of behaviour. One-zero scores are often highly correlated with both frequency and duration measures of the same behaviour, which means that they give a composite measure of 'amount' of behaviour. In contrast, frequency and duration measures of the same behaviour are not always highly correlated with one another. In some cases, therefore, a one-zero score may be a more meaningful index than frequency or duration.

Summary

Sampling may be done by focusing on a given individual, scanning a group of individuals or focusing on a behaviour pattern of particular interest. Recording may be done continuously, which is ideal but may be impractical or unreliable. Alternatively, recording may be done by time sampling. Two methods of time sampling involve measuring what is happening at particular moments (instantaneous sampling) or recording whether or not a given event has occurred within a given period (one-zero). Both techniques are easier to use than continuous recording but, without taking certain precautions, they may introduce inaccuracies.

6

The recording medium

The options available

Having looked at the various forms of behavioural measure, we now move on to consider the mechanical processes involved in recording them. The choice of the medium, or physical means, used to record behavioural observations has important consequences for the sorts of data that can be collected and the sampling techniques that can be used. Five basic methods of recording behaviour are available: video recording; written or dictated verbal descriptions; automatic recording devices; paper check sheets; and computer event recorders. The most flexible and commonly used methods are check sheets and computer event recorders. Using check sheets, an event recorder or any other method obviously pre-supposes that you have formulated a set of discrete behavioural categories.

Video recordings give an exact visual (and perhaps audio) record of the behaviour, which can subsequently be slowed down for analysis. Such date-stamped evidence is sometimes used as proof that the observer saw what was claimed and may be needed for inspection by others. Video recordings are useful for studying behaviour that is too fast or too complex to analyse in real time. Similarly, exact records of vocalisations can be made with an audio recorder and the sound patterns analysed later using specialised software. Digital technologies have largely superseded analogue methods. It is worth remembering, however, that when video recordings are stored on a computer they may be compressed and thereby lose quality.

The major advantage of recording behaviour with video is that the record can be analysed repeatedly and in different ways. The record also provides an accurate method of timing behaviour when a time code has been stored. It can be used for measuring inter-observer reliability.

Video recordings are rarely a complete replacement for direct observation however. At some stage it is always necessary to **code** the behaviour – that is, transcribe the record into quantitative measurements relating to specific behavioural categories. This may be done by recording observations relating to specific categories of behaviour on to a check sheet or computer event recorder while watching the recording.

Clearly, it is far more efficient (if possible) to code the behaviour as it actually happens and not postpone the measurement process by recording the behaviour instead. Behaviour is often easier to observe and analyse live and in context, rather than by watching it later on a screen, unless the behaviour is very fast or complex or involves many individuals. Thus, video should normally be used only for a specific reason – for example, if the behaviour requires repeated or very detailed analysis. If the behaviour is rare or of particular interest, permanent recordings are a useful back-up to live observation, ensuring that nothing is lost. Video recordings can also enliven oral presentations and bring to life aspects of behaviour that would be more difficult to demonstrate with words or static pictures (see Chapter 11).

A major drawback with video analysis is that it can be exceedingly time-consuming: behaviour that perhaps lasted only a few minutes may take hours to analyse. Moreover, the ease with which a recording can be replayed may make it tempting to analyse the record repeatedly and in ever greater detail, whereas it might be better to use this time for collecting more data. Another potential drawback of relying on a recording is the restricted field of view. Key details are sometimes lost.

Verbal descriptions of behaviour can be recorded in the form of long-hand written notes or dictated into a miniature audio recorder, which can be particularly convenient when recording outdoors in bad weather. Verbal descriptions are especially valuable during informal pilot observations and for recording rare events that are not a central feature of the study. In addition, dictating a verbal description is sometimes necessary for recording complex behaviour involving several categories that cannot reliably be coded directly on to a keyboard or check sheet.

As with video, spoken or written records must at some stage be coded if they are to be quantified, and transcription can be similarly time-consuming. Some researchers report that transcribing verbal records may take 10 to 15 times as long as the original observation time.

Automatic recording devices can be used to measure some kinds of behaviour. It is relatively simple to generate an electrical signal indicating when a lever is pressed, when an animal stands on a particular area of the floor, when it vocalises, or when it moves. An enormous range of devices is available, many giving a digital record suitable for storage and analysis on a computer. Efforts are being made to use pattern-recognition devices to measure the performance of complex activities (for example see www.noldus.com). Undoubtedly these devices and the accompanying software will continue to improve and when planning a study you could do well to discover how far the technology has advanced.

Activity monitors can record movements automatically by means of infrared, ultrasonic, capacitative or microwave detectors, the interruption of optical or infrared light beams, or the disturbance of sensitive microswitches. Physiological measures, such as heart rate, rectal temperature, blood pressure, electromyogram (EMG), or electroencephalogram (EEG) can be recorded remotely from sensors or electrodes using biotelemetry techniques. In the field, global positioning systems (GPS) are regularly used to record the position of an animal or the observer and a GPS is sometimes incorporated into event recorders.

A computer can be used both to run an experiment – for example, by turning equipment on and off, administering rewards, or opening and closing food hoppers – and to record the results; for example, by counting lever-presses or revolutions of a running-wheel, or by monitoring data from physiological sensors. Sensors and other measurement devices can also be connected to a computer. The list of possibilities is enormous and is limited mainly by the ingenuity of the experimenter. Our primary concern in this book, however, is with observational measurement techniques and we should stress that even in the most exquisitely and comprehensively automated study, in which virtually everything has been wired up, it is still essential to *watch* what is going on.

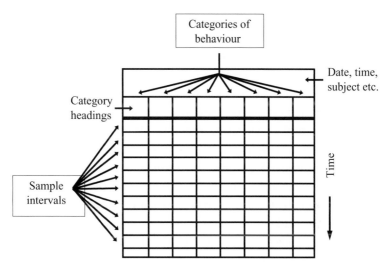

Figure 6.1 The basic design of a check sheet. Columns denote different categories of behaviour and rows denote successive sample intervals. It is important to include standard information such as the date, time, subject, place and so forth at the top of the sheet.

Check sheets

A check sheet is a simple paper-and-pencil tool that provides a cheap, flexible and surprisingly powerful way of recording observations. With practice and a correctly designed check sheet a considerable amount of information can be recorded reliably and with reasonable accuracy. It would be a mistake to assume that check sheets are universally inferior to technically more glamorous solutions.

The basic design of a check sheet is a grid, with *columns* denoting different categories of behaviour and *rows* denoting successive sample intervals (see Fig. 6.1). At the end of each sample interval, on the signal from a timer, you start recording on the next row of the check sheet. The choice of sample interval (the time represented by each successive row on the check sheet) was discussed in Chapter 5. Columns denoting related categories of behaviour, or behaviour patterns that tend to occur together, should be grouped on the check sheet.

It is always a good idea to leave one or two blank columns for additional information or new categories that may be added during preliminary observations. Another helpful feature is a remarks column, which can be used to note down unexpected events or incidental information that may later be useful in interpreting the results.

Some field studies combine focal, scan and ad libitum sampling with different parts of the same check sheet set aside for categories recorded using the different sampling rules. In addition, all three recording rules described in Chapter 5 (continuous recording, one-zero sampling and instantaneous sampling) can be used on the same check sheet for recording different categories of behaviour. With instantaneous sampling, behaviour is recorded by marking 'on the line' at each sample point. With one-zero sampling, a mark is made 'in the box' when the behaviour pattern first occurs. For continuous recording of frequencies, each occurrence is marked down within the appropriate sample interval. The exact frequencies of events as well as their sequence can be recorded relatively easily on a check sheet.

Recording precise durations on a check sheet is more difficult, since the actual times at which the behaviour started and stopped must be written down from a timer (although skilled observers can learn to do this quickly and accurately). A cruder way of recording approximate durations is simply to mark within the appropriate sample interval (the current row) whenever the behaviour stops and starts. This is obviously easier and quicker than reading a timer and writing down the actual times, but timing is only accurate to the nearest sample interval. If the sample interval is very short relative to the average bout length then this method may be adequate for many purposes.

An experienced observer can learn to record behaviour while only occasionally glancing down at the check sheet. Some observers have found that placing a grid over the check sheet allows the pencil to be guided to the right position by touch, obviating the need to look down; one hand is used to locate the correct square on the grid, the other hand to write.

The number of columns on a check sheet will obviously depend on the number of categories recorded, but it can be reduced by using different symbols within each column to represent different categories or sub-categories of behaviour. For example, within a single column labelled 'vocalisations', different symbols can be used to record particular types of vocalisation. Symbols can also be used to identify individual animals, such as the initiator and recipient of a social interaction. If too many columns are used, however, recording will become difficult and therefore inaccurate.

If data from check sheets are going to be analysed using a computer, then the design of the check sheet should take this into account by making it easy for the results to be transcribed directly on to a keyboard in a suitable format. A check sheet design that necessitates a two-stage transfer of data, from check sheet to a coding sheet and thence on to a computer, wastes time and creates additional opportunities for error. Bear in mind, though, that data should be easily accessible for exploratory data analysis (see Chapter 9).

Event recorders

Basic principles

A computer event recorder is simply a portable or hand-held computer or personal digital assistant (PDA) with the appropriate software that enables you to record behavioural observations directly on to the machine.

For most applications all that is needed is an ordinary computer or PDA and suitable event-recording software for encoding and interpreting the data inputs. With the right software, virtually any device will do.

A number of excellent software packages for transforming standard lap-top or hand-held computers into behavioural event recorders are commercially available. Some of these packages also include software for analysing and presenting the data.

Behavioural observations are normally recorded as key-presses, either on the computer's standard keyboard or touch-sensitive screen or, less commonly, on a specialised keyboard or other input device. Each key denotes a particular category of behaviour or a particular subject. Every time a key is pressed the computer records the identity of the key, together with the time at which it was pressed. As a further sophistication, it may be possible to use some keys as 'modifier' keys – for example, to denote the quality or extent of a behaviour pattern, or the initiator or recipient of a social interaction. Trainable voice-recognition software is also available, enabling you to communicate with the computer using speech commands rather than via the keyboard.

The simplest type of event recorder software is capable of recording frequencies or durations for several categories of behaviour, but it may not be able to deal adequately with the complexities that arise when several individuals are watched simultaneously. In studies of social behaviour it is often necessary to record interactions of the type 'Subject/Verb/Object' – for example, 'Animal A/approaches/Animal B' or 'Mother 2/nurses/Pups 1,3 & 5'. This type of event recording requires somewhat more sophisticated software for analysis (though not necessarily for recording). One way round this problem is to use the event recorder as a check sheet emulator. In this case, information is recorded at regular intervals (sample intervals) in the form of a string of key-presses, the final key-press of each string denoting the timing. The advantage of this method is that complex behaviour, such as social interactions, can be recorded using only simple event-recording software.

Computer event recorders offer at least four obvious advantages over paper check sheets:

- Durations can be recorded with much greater precision.
- You can record more rapidly-occurring streams of behaviour and use more categories.
- They eliminate the laborious and error-prone task of transcribing raw data from check sheets into numerical form, which can sometimes take longer than collecting the data in the first place.
- Large quantities of data can be stored in a much more compact form than is possible with check sheets – a factor that is especially important for field workers.

However, event recorders also have certain potential pitfalls:

- They offer a degree of sophistication that may be unnecessary for some applications. There is no point in using a complex piece of technology if paper and pencil would do just as well. As a general rule, it is wise to choose the simplest recording technique that will do the job.

- Event recorders are not always as readily adaptable as check sheets, which can be designed specifically by the user to suit each application. The recording medium should be designed around the behavioural categories to be recorded, and not vice versa.
- The ease with which data can be collected using an event recorder can be a mixed blessing, since it is possible to collect too much information as well as too little. Using an event recorder should not be an excuse for being vague about the questions being asked or collecting irrelevant data.

Essential and desirable features for an event recorder

Any event recorder, whether it is a standard computer adapted for that purpose or a custom-built machine, should have the following general features:

- The keyboard or input device should have a sufficient number of keys or input combinations to record all the categories, preferably without using multiple key-presses (such as SHIFTed keys). In the rare cases where the number of keys of a conventional keyboard is insufficient for very complex behavioural observations, a custom-built keyboard with more keys may have to be used.
- Visual feedback should be presented on a screen to confirm that the correct data have been entered. The screen can also provide you with additional information, such as real or elapsed time and an indication of key states ('on' versus 'off' for duration keys).
- It should be easy to rectify errors quickly by deleting mistaken key-presses. Good event-recording software may have the facility to detect and correct obvious keying errors automatically.

The quality and reliability of the hardware and software in most modern hand-held computers are quite adequate for all but the most exacting applications. Waterproof devices are available (at a price) for those who really need them for underwater observation, in rainforests or other wet habitats.

The following additional features are desirable, if not always necessary, in an event recorder:

- It should be easy for the user to customise the keyboard or input device by selecting keys to perform particular roles, allowing the machine to be used for more than one type of observation protocol. Some software allows the user to assign each key to a particular role by asking a series of questions when the machine is initially set up.
- Nearly all event recording devices allow the user to transfer data to another computer for permanent storage and detailed analysis.
- It is helpful if you can also record plain-text notes or comments, in a similar manner to the 'remarks' column on a check sheet. (Even if this is possible, it is still a good idea to have a notebook or voice recorder handy just in case.)

A computer can also be used to record data automatically from one or more inputs, switches, cameras or measurement devices such as sensors fixed on or in the subject or activity monitors (see for example Kahng & Iwata, 1998). Some companies are specialising in building devices for the automatic collection of behaviour (e.g. NewBehavior AG: www.newbehavior.com)

Commercially available event recorders are readily available and are becoming increasingly linked to pattern-recognition devices that distinguish between different patterns of behaviour. The recorders and their associated software products are not always cheap but many laboratories have found them so useful that they have invested heavily in them. Noldus Information Technology (see www.noldus.com) has been the market pioneer in developing many of these products for behavioural studies under the trade names of The Observer and Ethovision, but other firms have entered the market so it is worth shopping around (see for example Interact by Mangold Software & Consulting: www.mangold.de).

Summary

Recording techniques range from simple pencil and paper to computer automation. The most flexible are check sheets and computer-based event recorders. Recording behaviour with video has certain advantages but usually means that a lot of work must be done before measurement can start. Computer-based systems also have their disadvantages but they can enormously reduce the time taken in analysis and can combine direct recording by the observer with recording by automatic devices.

7

How good are your measures?

Having decided which aspects of behaviour to measure and chosen the suitable recording medium, you would be wise to check the quality of your measurements before you proceed to collect a lot of data. Measuring behaviour, like measuring anything else, can be done well or badly. When assessing how well behaviour is measured, two basic issues must be considered: **reliability** and **validity** – sometimes expressed as the distinction between 'good' measures and 'right' measures.

Reliability versus validity

Reliability concerns the extent to which measurement is repeatable and consistent: that is, free from random errors. An unbiased measurement consists of two parts: a systematic component, representing the true value of the variable, and a random component arising from imperfections in the measurement process. The smaller the error component, the more reliable the measurement. Reliable or good measures are those that measure a variable precisely and consistently.

At least four related factors determine how 'good' a measure is:

- *Precision*: How free are the measurements from random errors? The degree of precision is represented by the number of significant figures in the measurement. Note that precision and *accuracy* are not synonymous: accuracy concerns systematic error (bias) and can therefore be regarded as an aspect of validity (see below). A clock may tell the time with great precision (to within a millisecond), yet be inaccurate because it is set to the wrong time.

- *Sensitivity*: Are small changes in the true value invariably reflected by changes in the measured value?
- *Resolution*: What is the smallest change in the true value that can be detected?
- *Consistency*: Do repeated measurements of the same thing produce the same results?

If behavioural characteristics are measured unreliably then real effects, such as differences between groups or correlations between measures, may remain undetected.

Validity concerns the extent to which a measurement actually measures what the investigator wishes to measure and provides information that is relevant to the questions being asked. Validity refers to the relation between a variable, such as a measure of behaviour, and what it is supposed to measure or predict about the world. Valid measures, sometimes referred to as right measures, are those that actually answer the questions being asked.

To decide whether a measure is valid, at least three separate points must be considered:

- *Accuracy*: Is the measurement process unbiased, such that measured values correspond with the true values? Measurements are accurate if they are relatively free from *systematic* errors (whereas *precise* measurements are relatively free from *random* errors).
- *Specificity*: To what extent does the measure describe what it is supposed to describe, and nothing else?
- *Scientific validity*: To what extent does the measurement process reflect the phenomena being studied and the particular questions being asked? Do the measurements tell you anything important? Ultimately this is a matter of scientific judgement and encompasses issues such as whether, for example, data from laboratory experiments on a particular species may be generalised to other species or to animals living in natural environments.

Suppose, for example, that you wish to assess how much milk a young mammal receives from its mother and that you use a behavioural measure – the total duration of suckling – to assess this. The behavioural measure is only valid for this purpose if the total duration of suckling is strongly correlated with the amount of milk transferred. In some species, such as rats and pigs, the relationship between suckling duration and milk intake is weak. Young piglets and rat pups spend a lot of time suckling but obtain no milk for most of the time they suckle. Milk release only occurs during brief periods (less than a minute), which are separated by long intervals (20 min or more). Therefore suckling duration is not a valid measure for assessing milk transfer in these species, despite the fact that it can be measured precisely and consistently. It is not the *right* measure even though it may be a *good* measure.

It is relatively easy to devise behavioural categories, tests, questionnaires or interview techniques that are believed to measure some aspect of behaviour, personality or intelligence, but it requires external evidence (validation) to demonstrate that they measure what they are supposed to measure. Highly reliable (good) results can be obtained using biased, irrelevant or meaningless (wrong) measures. Methodological rigour must be sacrificed sometimes in the interests of measuring the things that really matter. It may be better to measure the right thing imperfectly rather than the wrong thing with great precision. The point can be made by reference to shots on a target (Fig. 7.1). Four possibilities are shown: imprecise and inaccurate, precise but inaccurate, imprecise but accurate, and precise and accurate. The last is most desirable but it is better to be imprecise and accurate than precise and inaccurate.

In addition to the questions of reliability and validity, an important practical consideration is **feasibility**. This concerns the extent to which the proposed measurement procedure is possible, practicable and worthwhile. Does the information obtained justify the cost and effort required and the disruption to the subjects?

Within-observer versus between-observer reliability

Observers can be regarded as instruments for measuring behaviour in much the same way that, say, a thermometer is used to measure temperature. Just as measuring instruments may be biased or imprecise, so too may errors in measuring behaviour arise from variations

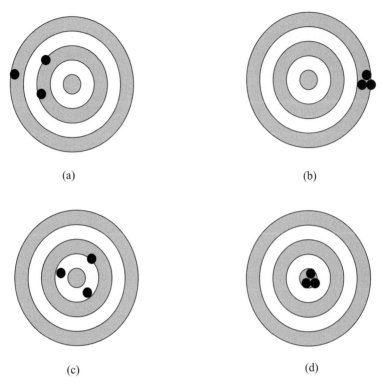

Figure 7.1 Targets have been hit in four ways. The shooting was (a) imprecise and inaccurate, (b) precise but inaccurate, (c) imprecise but accurate, and (d) precise and accurate. In behavioural studies it is always worthwhile being accurate but sometimes it may be necessary to lose some precision in order to achieve it.

within or between observers. Two different aspects of reliability can be distinguished: within-observer reliability and between-observer reliability.

Within-observer reliability (or observer consistency) describes the extent to which a single observer obtains consistent results when measuring the same thing on different occasions. To assess within-observer reliability, the observer measures exactly the same sample of behaviour on two or more separate occasions. Typically, this involves videoing the behaviour. Ideally, you should conduct tests of your reliability at regular intervals during the course of a study, especially if it is a lengthy one.

Between-observer reliability describes the extent to which two or more observers obtain similar results when measuring the same thing simultaneously. This is a measure of the agreement between different observers attempting to measure the same thing. To assess between-observer reliability, a sample of the behaviour is measured simultaneously by two or more observers, either live or from a video recording.

In any study involving two or more observers, you should verify first, that each observer consistently records in the same way on different occasions (i.e. that each observer exhibits good within-observer reliability); and second, that the observers are recording the same behaviour in the same way (i.e. that the between-observer reliability for each category of behaviour is good).

Establishing good agreement between observers is especially important if, as sometimes happens, the observations on one group of subjects are made mainly by one observer while those on the comparison group(s) are made by a different observer. Clearly, the danger here is that apparent differences between the groups could stem entirely from differences between the observers.

Even in studies involving only one observer, a demonstration of good *between*-observer reliability can still be valuable. Doing this obviously requires finding a second observer for some of the recording sessions. The point of assessing between-observer reliability in a single-observer study is that the principal observer might unknowingly be highly reliable at measuring the *wrong* behaviour (wrong, that is, with respect to the stated definitions). Good within-observer reliability demonstrates internal consistency, but it does not demonstrate that another observer would record the same behaviour. Thus, even when all measurements are made by one observer it is still helpful to demonstrate that a second observer would have produced a similar set of results using the same methods.

Measuring reliability using correlations

Reliability is often and most simply expressed as a correlation coefficient (r) (see Chapter 9). A correlation of $+1.0$ denotes a perfect positive association between two sets of measurements, while a correlation of

Table 7.1 *Inter-observer reliability*

Session	1	2	3	4	5	6	7
Frequency (h^{-1})							
Observer A:	23	12	34	17	24	13	37
Observer B:	18	15	30	22	25	10	41

zero denotes a complete absence of any linear association. A reliability correlation can be calculated for each measure or category of behaviour.

To score within-observer reliability, the observer measures each of *n* independent samples of behaviour (for example, *n* different video recordings) on each of two separate occasions. The reliability for each category of behaviour is then calculated as the correlation between the *n* pairs of scores.

To score between-observer reliability, two observers simultaneously measure each of *n* independent samples of behaviour (for example, during *n* different recording sessions). Again, the reliability of each category is calculated as the correlation between the *n* pairs of scores. When quoting reliability, the results should state whether the correlations are Pearson or Spearman correlation coefficients and give the number of pairs of scores (*n*) on which the correlations are based.

Measuring reliability requires *independent* pairs of measurements. It is not correct to base a reliability correlation on a single sample of behaviour – for example, one observation session or video recording – which has been split up into several short sub-samples, since the measurements would not be independent of one another. In addition, the test samples used to assess reliability should be fair (and preferably random) samples of the behaviour that is actually measured in the study. Otherwise, it would be easy to obtain a spuriously high reliability by choosing to analyse a sample in which, say, the behaviour never occurred or occurred continuously. Ideally, you should choose the test samples randomly and measure reliability under the normal conditions of the study.

A simple illustration of inter-observer reliability is shown above in Table 7.1. Two observers (A and B) each recorded the frequency of the same behaviour pattern during seven different recording sessions.

The inter-observer reliability expressed as a Pearson correlation coefficient (r) is $+\,0.92$ ($n = 7$; 5 d.f.).

How reliable is reliable?

No magic figure exists, above which all measures are acceptable and below which none are. Acceptability depends on several factors, including the importance of the behavioural category and the ease with which it can be measured. In the case of an important category that is difficult to measure, a rough guideline for acceptability might be a correlation of at least 0.7. The reasoning behind this criterion is that with a correlation coefficient of 0.7 roughly 50% of the variance in one set of scores is accounted for statistically by the other set of scores, since the coefficient of determination $r^2 = 0.7^2 = 0.49$ (see Chapter 9). We must stress, though, that this is only an informal guideline; some authorities might argue that a reliability of 0.7 is too low for any measure, no matter how important. For categories of behaviour where measurement is straightforward, reliability should be well above 0.7.

Finally, be aware that the level of statistical significance (the 'p value') of the correlation says little about the degree of reliability, because the level of statistical significance attached to a correlation depends on the sample size as well as the strength of the association or 'effect size' (see Chapter 11 on the important distinction between effect size and statistical significance). The *size* of the correlation coefficient is what matters, not its statistical significance. For example, a correlation coefficient of 0.5 would signify a poor degree of reliability, yet this correlation would be highly statistically significant ($p < 0.01$) with a sufficiently large sample size ($n > 26$ pairs).

Other ways of measuring reliability

Among the various other indices of observer reliability, three common and relatively simple measures are the index of concordance, the kappa coefficient and the Kendall coefficient of concordance.

Index of concordance

A method for assessing reliability which is particularly suited to nominal or classificatory measures is to note whether there is categorical (yes-or-no) agreement about each occurrence of the behaviour. At the end of an observation session, the two observers compare the total number of agreements (A) and disagreements (D). The index of concordance is the proportion of all occurrences about which the two observers agreed, i.e. $A/(A + D)$. The same index expressed as a percentage is sometimes referred to as the percentage agreement.

An index of concordance, rather than a correlation, need only be used if there is some reason why categorical agreement over each occurrence of the behaviour is an important issue, or if the behaviour is measured on a nominal scale. As a general rule, an index of reliability should be calculated using the same type of measure – such as a frequency or total duration – as is used in the final presentation and analysis of the results.

Kappa coefficient

The simple index of concordance described above does not take account of agreements that arise by chance alone. The kappa coefficient is an index of inter-observer reliability which does allow for chance agreements and is given by $(O - C)/(1 - C)$, where $O =$ the observed proportion of agreements (i.e. the index of concordance, as defined above) and $C =$ the proportion of agreements that can be accounted for by chance. For example, suppose two observers simultaneously recorded a behaviour pattern for 15 min, using instantaneous time sampling at 30-s intervals (i.e. with 30 sample points). Observer 1 recorded the behaviour on a total of 16 sample points, while observer 2 recorded the behaviour on 13 sample points. The two observers agreed (A) about the occurrence or non-occurrence of the behaviour on 25 sample points and disagreed (D) on 5 sample points. The observed proportion of agreements $O = A/(A + D)$ $= 25/30 = 0.83$. The chance proportion of agreements (C) is given by the probability that the observers will both score an occurrence (or both score a non-occurrence) on the same sample point if their scores are distributed randomly, i.e. $C = (16/30 \times 13/30) + (14/30 \times 17/30) = 0.23 + 0.26 = 0.49$. Therefore, the kappa coefficient $\kappa = (0.83 - 0.49)/(1 - 0.49) = 0.34/0.51 = 0.67$. As expected, the kappa coefficient is considerably lower than the simple index of concordance, showing that a substantial number of agreements may be accounted for by chance alone.

Kendall coefficient of concordance (W)

In studies where three or more observers measure the same behaviour, the Kendall coefficient of concordance (W) can be used to quantify the overall agreement among them. The coefficient W is a non-parametric statistic which expresses the degree of association among sets of rankings (details can be found in Siegel & Castellan, 1988).

Factors affecting reliability

Many different factors affect how well a category of behaviour is measured, and these relate both to the measurement technique and the nature of the behaviour itself. Among the most important factors influencing reliability are the following.

Practice and experience

Quite simply, the more experienced you are and the more you have practised the measurement techniques, the better will be the results.

Frequency of occurrence

If a behaviour pattern occurs very rapidly it may be difficult to record each occurrence reliably. Little can be done about this, except to record the behaviour on video and analyse it later in a slowed-down form. Conversely, rare behaviour patterns may be missed altogether if observation sessions are not long enough.

Observer fatigue

If an observation session lasts too long, your ability to record accurately may be impaired through fatigue and loss of concentration. You must balance the quantity of data collected in each session against the quality of the data.

Adequacy of definition

One potential source of trouble that can often be improved is how well each category of behaviour is defined. If a category is not clearly and unambiguously defined then it probably cannot be recorded reliably.

A common problem with protracted studies is that definitions and criteria tend to drift with the passage of time, as observers become more familiar with the behaviour and, perhaps unconsciously, 'improve' or 'sharpen up' the definitions. This phenomenon is referred to as **observer drift**. One way of guarding against observer drift is to measure reliability at the beginning, the end, and at various points during the study. The simplest precaution, though, is to write down the clearest possible definitions and ensure that all observers are fully familiar with these throughout the study.

Dealing with unreliable measures

What happens if an important measure – one that is central to the question being asked – turns out to be unreliable? If statistical purity were the only arbiter then all unreliable measures would automatically be discarded. This should certainly be the fate of any unreliable measures that are also irrelevant or uninformative. Nonetheless, unreliable ('bad') measures are sometimes important ('right') and therefore you may have to sacrifice statistical purity for the sake of measuring what really matters.

Two things can be done to help alleviate the problem of unreliable measures. First, unreliable measures can sometimes be redefined, or the measurement technique improved, to make them more reliable. Modifying the definition, perhaps by making it more restrictive in scope, may eliminate ambiguous cases that are difficult to categorise. Obviously, data acquired after a definition is changed cannot be combined with data acquired before the change. If a category is difficult to record reliably using continuous recording, time sampling may give a more reliable record. For some difficult categories, reliable measurement just requires lots of practice.

A second possible approach is to combine two or more unreliable measures to see whether together they produce a more reliable composite measure. Combining related measures also eliminates redundancy and reduces the number of categories used in the final analysis and presentation of results. When the absolute frequencies of the separate scores are generally low and many individuals have scores of zero, the combined score may be more sensitive than any of its elements and easier to analyse statistically. For example each component of grooming might have occurred infrequently but, when combined for each animal, the overall scores might be higher.

If measures are combined in this way, the composite measure must have 'face validity': that is, the separate measures must mean the same and the composite measure must make intuitive and biological sense. The separate measures must therefore describe causally and functionally similar behaviour patterns.

How independent are the measures?

Statistical tests (described in Chapter 9) generally assume that the data consist of a random sample from the population and that individual data points are statistically independent of one another. This basic assumption of independence is sometimes violated in behavioural research.

The pooling fallacy

A common error is to treat repeated measurements of the same subject as though they were independent of one another. Suppose you wished to estimate the average stride length of adult male giraffes and the sample of individuals available was necessarily small. In the mistaken belief that you were increasing the sample size, you might measure as many strides as possible for each giraffe. Eventually you might acquire 100 measurements ($k = 100$) for each of ten adult male giraffes ($n = 10$). If you pooled them to give 1000 measurements ($'N' = nk = 10 \times 100$), you would have fallen foul of the pooling fallacy. The true sample size (n) is ten, not 1000. The error becomes more obvious if we substitute, say, weight for stride length in this example. Weighing ten individuals 100 times each and treating the sample size as though it were 1000 rather than ten would clearly be wrong.

The crucial point is that obtaining additional measurements from the same subjects is not a substitute for increasing the number of subjects in the sample. In general, repeated measurements from an individual should be averaged to give a single data point for that subject and the sample size (n) should be equal to the number of *subjects*, not the number of measurements.

It is not necessary to throw away information about variation for a single individual, particularly when all individuals have been measured approximately the same number of times. Statistical methods such as repeated measures Analysis of Variance (ANOVA), which we describe in Chapter 9, can be used to analyse within-subject variation as well as sources of variation between subjects. The relative benefits of increasing the sample size over increasing the time devoted to each subject are discussed in Chapter 8.

Group effects

Problems with statistical independence also arise when studying species that give birth to multiple offspring – for example, rats or cats – or when animals are housed together in groups. The difficulty here is that individuals within the same litter or group may be more (or less) similar to each other than they are to individuals in other litters or groups. Animals housed in the same cage may be more alike than animals from other cages. Similarly, in studies of social behaviour, interactions between members of the same group can lead to problems of independence, through short-term effects on their behaviour, or through less obvious shared factors such as eating locally available foods or being exposed to local predator pressure.

If within-group variation is significantly smaller (or larger) than between-group variation, measurements of group-mates cannot be considered to be statistically independent. For example, suppose you are interested in the effect of an experimental treatment on the weight of rat pups and assign six pups from the same litter to be the experimental group, and six pups from another litter to be the control group. If, by chance, the mother of the experimental group happened to be large then her pups would also tend to be large, thereby confounding a litter effect with the treatment effect. The sample size of each group should be one (litter) and not six (pups).

One solution to the problem of group effects is to measure only one randomly selected individual from each group, although this procedure throws away potentially valuable information. A better alternative is to measure all the group-mates, but to treat the *group-mean* value as a single measurement, so that the sample size (n) is equal to the number of groups rather than the number of individuals. A third approach is to use nested Analysis of Variance to pick out litter or group effects (see Chapter 9). This may require a balanced experimental design, in which one or more members of each litter or group (preferably equal numbers) are included in each treatment group, so that the effects of the experimental treatment and between-litter variation are not confounded.

Sometimes the variation within litters or groups is *greater* than the variation between litters or groups. In other words, group-mates can be *less* alike rather than more alike. This increase in within-group variation can arise if social interactions – for example, competition – accentuate differences between animals that are housed together. Considerable evidence suggests that human siblings are less like each other in terms of certain behavioural characteristics than would be expected on a chance basis (Dunn & Plomin, 1990). Such phenomena can be uncovered with the appropriate use of Analysis of Variance.

Non-independent categories

When using different types of measure to assess the same behaviour pattern, it is important to ensure that the various measures are, in principle, independent of one another. Suppose, for example, that a particular type of behaviour is measured in three ways: in terms of its mean duration, total number of occurrences, and total duration. Now, these three measures are not independent, since mean duration is equal to the total duration divided by the number of occurrences. Given any two of these measures, the third can be calculated, so only two measures can be regarded as independent descriptions of the behaviour.

Non-independence can cause problems in interpreting associations between measures. Two measures may be correlated either because the two types of behaviour are associated for non-trivial reasons or because the two measures are merely two different ways of measuring the same thing. For instance, suppose two mutually exclusive categories of behaviour are found to be negatively correlated. If one category, let us say 'sleeping', automatically excludes a second category of 'awake', then 'awake' is merely equivalent to 'not-sleeping'. Similarly, if a wild herbivore spends most of its waking time grazing, then a measure of time spent grazing and a measure of time spent sleeping are bound to be negatively correlated.

Summary

Measurements can be made precisely but without being valid. In behavioural observation, reliability is important and a variety of techniques are available to check that different observers record the same things. Valid measures answer the question that is being asked. Questionable independence of measurements may arise if an observer has recorded many times from the same individual and each measure is treated as equivalent to measures from other individuals. Measures from members of the same group must be treated with care since they may not be as independent of each other as measures taken from individuals in other groups. Also measures may not be independent of each other if they occur frequently and are mutually exclusive.

8

How good is your research design?

Formal prescriptions about how scientific research should be conducted often fail to capture the creativity of the best scientists. Therefore, advice on how to design research must be given and taken with caution. Some research might simply involve carefully watching an animal to see what it does next. This type of work should not be scorned. Performing an experiment may seem more 'scientific' than open-ended observation but the yield may be less. Many questions about behaviour are most appropriately answered by non-experimental observational research. Such work can also help to distinguish between alternative explanations if, for example, naturally occurring events demonstrate associations between variables that previously seemed unrelated, or break associations between variables that previously seemed to be bound together. Moreover, worthwhile experimental research almost invariably needs to be preceded by careful observation. Knowledge of the normal behaviour of animals, preferably in their natural environment, is an invaluable precursor to experimental research.

Performing experiments

The point of an experiment is to find out whether varying one condition produces a particular outcome, thereby reducing the number of plausible alternative hypotheses that could be used to account for the results. You will almost inevitably have some expectations about the outcome of an experiment, even if you are not consciously aware of these expectations. This potential source of bias can be removed by ensuring that the person making the measurements is unaware of which treatment each subject has received until after the experiment is over. This procedure is referred to as running a **blind** experiment.

You may influence your own observations in the direction of a favoured hypothesis. The cumulative effect of many such minor biases may be an apparent significant difference between the experimental and control groups. These effects are often surprisingly large and the only sure way to minimise them is for you to be genuinely unaware of how the subjects have been treated until after the data have been collected.

If you are not blind to the treatment when the measurements are made, you may also unconsciously provide the subjects with clues that influence their behaviour in a particular way. The most famous example of this was the case of Clever Hans, a performing horse that seemed able to count. Only as a result of careful experiments under blind conditions was it found that the horse was not able to count but was, in fact, responding to subtle cues unconsciously produced by its trainer. Be aware of how you may unintentionally affect the subjects' behaviour and bias scores in the expected direction.

If the subjects of a study are humans, they too may introduce bias into the results if they are aware of the group they are in or the treatment they have received. If at all possible, and ethically acceptable, human subjects should not be aware of which group they are in until after the study is over. An experiment in which neither the person making the measurements, nor the subjects, know the treatment that each subject has received is called a **double blind** experiment. This type of experimental design is widely used, for example, in assessing the clinical effects of drugs and other forms of medical intervention.

Experimental design

Careful attention to the design of experiments will save many a headache at a later stage in a study. Prior discussion with a statistician or an experienced colleague is usually time well spent. A statistician will be able to advise on efficient research and how to code the data so that they fit easily with suitable software. Data handling skills and statistical knowledge can play a vital role in getting the most out of a project.

Good design is efficient, saving both time and money by winning greater precision and reducing the numbers of subjects required in order to answer a question. Also, good design can reveal interactions between factors that would never have been spotted if you had supposed that examining one factor at a time was sufficient.

Understanding how to get the most out of an experiment involves knowing what analytical tools are available, and this chapter should be read in conjunction with the next one. If information such as the sex, weight, habitat and history of each animal is collected at the time of the experiment, techniques are available to improve the precision of the analysis and extract much more from the data than if such collection of information had been neglected. Many excellent textbooks have been written about experimental design (Mead, 1988; Ruxton & Colegrave, 2003) and here we merely summarise some of the main points and highlight issues that are likely to arise in behavioural work.

The aim of the simplest experiment is to vary just one condition (the **independent variable**) and measure the effect on one or more **dependent variables** (or outcome measures), while holding all other conditions constant. The effects of varying the condition (the so-called **treatment effect**) are measured for one group of subjects (the **experimental** or treatment group). They are compared with those for a **control** group of subjects which do not receive the experimental treatment but are – at least in theory – similar in all other respects. To achieve this, the subjects must be assigned randomly to the experimental and control groups and treated identically, except with respect to the factor in which you are interested.

For example, you might wish to investigate whether a bird inherited the behavioural characteristics of its parents by placing the parents' eggs in the nest of unrelated individuals (known as cross-fostering; see below). Failure to find any similarity with the natural parent might arise from **confounding factors** such as the disruption of moving eggs around. So an appropriate control group would be one in which the eggs were removed for a while and then returned to the parental nest.

In most experiments it is difficult to vary one condition without varying something else as well. Part of the art of good experimenting lies in picking the appropriate control groups and randomising or otherwise eliminating the effects of confounding variables. In these ways it becomes possible to distinguish between the effects of variables that would otherwise be entangled with each other. One standard procedure is known as **blocking** – a term that originally came from agricultural experiments and refers to blocks of land that were chosen so that conditions of the soil or position were not confounded with the experimenter's treatments. If you are forced to test animals at different times of the day and you want to eliminate the possibility that time of testing caused what might be revealed as differences between an experimental and a control group, you should ensure that equal numbers of individuals from both groups are tested in the morning and in the afternoon. More complicated experimental designs are described in many textbooks such as Mead (1988) and Ruxton and Colegrave (2003).

Any set of measurements obtained from a group of individuals will exhibit some variability. This variation may be owing partly to fluctuations in the environment (including the impact of humans on the subjects). The subjects themselves may be of different sexes or ages, they may be at different stages in their diurnal cycles or reproductive cycles, their health may differ or they may come from different environments or genetic backgrounds.

The type of treatment imposed on the subjects is likely to be the most obvious so-called 'fixed effect' on variation in the subjects' behaviour. Other fixed effects such as those owing to age or sex can be managed by ensuring that they are not confounded with the treatment. They may also be taken into account in subsequent statistical analysis.

So-called 'random effects' may arise from unintended or unwanted variation in the environment, or from the characteristics of the subjects, or from poor measurement techniques. The skilful experimenter will attempt to eliminate these sources of variation. For example, if rats or mice are suitable subjects for the study, genetically similar inbred strains will generally produce less variable responses than outbred strains. Note, however, that strains often differ markedly from each other so that what is gained by reduction in variation is lost by limitations on what can be generalised from the results.

Completely randomised designs involve straightforward comparisons between an experimental and a control group. They are used when the fixed effects, other than the treatment, are unlikely to be important. Such experiments are simple to conduct and easy to analyse. However, they take no account of factors that might interact in important ways with the experimental intervention. They would not, for example, indicate whether the treatment was more effective in one sex than in the other.

Factorial designs test simultaneously the potential effects of several factors such as the experimental treatment, the time of day and the sex of the subjects, together with the interactions between these factors. These designs are a much more efficient alternative to conducting several simple experiments. The effects of two conditions may not merely add together: they may **interact** so that a particular combination of conditions produces an effect that might have been missed if each condition had been varied one at a time and the others held constant. Some of the most illuminating experiments simultaneously examine two or more factors. If the point of such an experiment were to examine interactions between, say, sex and type of housing, then equal numbers of each sex would be exposed to each type of housing condition.

Results from factorial designs are usually analysed statistically using analysis of variance (ANOVA), which is described in the next chapter. However, the method of analysis and therefore the design may be rendered invalid if ranked or categorical measures are used (see Chapter 3 for different levels of measurement). Moreover, it is easy to make factorial experiments too elaborate. A balance must be struck between discovering what combinations of conditions do interact, and obtaining results that are so complicated that they are difficult to interpret.

Matched-pairs designs (also called within-subjects designs) are used because individuals can differ considerably from one another in their behaviour. It is therefore often desirable to remove the confounding effects of the constellation of factors that distinguish one individual from another. A simple way of doing this is to use each individual as its own control, by testing the same individual under both the experimental and control conditions on separate occasions. The scores are the *differences* between the experimental and control conditions for each subject. The sequence in which the experimental and control conditions are presented may also matter, so half the subjects should be presented with the experimental treatment first and the other half counter-balanced with the experimental treatment second. If more than two treatments are involved, the order effects inherent in repeatedly testing the same subject can sometimes be balanced between different groups, some individuals getting one order of presentation while others get another, so that every possible order is used. Alternatively, the order can be randomised, using what is called Latin square design described in many textbooks (e.g. Mead, 1988; Ruxton & Colegrave, 2003).

Another form of matching involves pairing each individual in the experimental group with an individual in the control group on the basis of one or more so-called matching variables. For instance, individuals might be matched on the basis of age, sex or body weight. The aim of matching pairs in this way is to reduce the effects of pre-existing differences between the two groups arising from differences in the effects of the matching variable(s). Such a matching design has two potential disadvantages, however. First, if individuals in the experimental and control groups are matched according to any criterion whatsoever, then for statistical analysis the sample size is effectively halved. Therefore, unless matching really does substantially reduce the initial variance in the behaviour being measured, it may actually result in a net *reduction* in the overall power of the statistical test. Second, the matched-pairs design can raise practical problems, since perfect matching on two or more variables might require an impractically large number of potential subjects. If, as is normally the case, only a limited pool of individuals is available to study, difficult choices may have to be made between dropping individuals from the study because no suitable partner can be found for matching, or relaxing the criteria for matching. Dropping individuals reduces sample size, and therefore statistical power, while relaxing the criteria reduces the effectiveness of the matching procedure. Either way, compromises may have to be made. In many cases this form of matching is more trouble than it is worth.

Repeated measures

Repeated testing of the same subject may markedly influence its behaviour: many processes such as arousal, sensitisation, conditioning, fatigue and habituation may contribute and interact so that the changes in the subject's responsiveness over time are not simple. These changes may themselves be of interest. If they are ignored, however, they can mean that measurements made in sequence are not comparable. The point to remember is that once an individual is tested, it becomes a somewhat different individual. A practical benefit of repeated measurement is that in many cases an individual does not have to be dropped completely from a study if one or two measurements are missed. The occasional, unavoidable gap in data collection can therefore be more easily tolerated. Analysis of Variance, described in the next chapter, can be used for analysing repeated measures.

Pseudo-replication

An otherwise well-planned experiment may still be limited in scope because only a restricted conclusion may be drawn from the results. You need to be sensitive to the subtle but pervasive problem that arises in both field and laboratory work, and which is sometimes known as the problem of **pseudo-replication**. A commonly given example is the following. While many birds of a given species may respond in the same way to one tape recording of song played to them, they might have responded differently to another tape recording that is thought to be the same but is subtly different (Kroodsma, 1989). This problem, which is discussed by Wiley (2003), is more general than one of finding a range of stimuli to present to animals. In any experiment the conditions to which a group of animals has been exposed may not be the same as another set of conditions that is supposedly the same.

Studying development

One of Tinbergen's four questions described in Chapter 1 is concerned with development or ontogeny, namely the changes that occur as an individual matures and the processes involved in such changes. Measuring the behaviour of young animals and children raises some special problems because the organisation of their behaviour alters as they develop. Activities that may look the same at different ages may be controlled in different ways and have different functions. For instance, the amount of time a young monkey spends in contact with the ventral surface of its mother is influenced primarily by its need for milk early in life, and later by its need for a refuge from danger. Some activities such as suckling in young mammals are special adaptations to an early phase and drop out of the repertoire as the individual becomes nutritionally independent of its mother.

Behavioural development can be studied by **cross-sectional** research, which involves measuring different individuals at each age, or by **longitudinal** research, which involves measuring the same individuals repeatedly over time.

Longitudinal measurement provides information about the manner in which each individual's behaviour changes as a function of time – that is, it provides information about the *process* of behavioural change as well as its outcome. However, age-related developmental changes and the consequences of experience of the test situation are inevitably confounded if the same subjects are repeatedly tested as they grow older. Repeatedly testing a young animal's responsiveness to a stimulus, for example, can influence the course of its development and thereby affect its behaviour in subsequent tests of responsiveness. Even the individual's increasing familiarity with the test situation can affect its behaviour in ways that are not related to any underlying developmental changes in, say, responsiveness. Thus, the general problem of order effects arising from repeated testing of the same subjects is particularly important in developmental studies.

Cross-sectional measurements are not vulnerable to order effects, but they do have a number of other drawbacks. First, individuals of two different ages may differ from each other in ways that are not merely the result of differences in age and experience between the two ages at which measurements are made. The two age groups may have differed in unknown ways prior to the age when the first group was measured. This could arise, for instance, if food availability to the parent had a marked influence on the overall pattern of development of its offspring (Gluckman & Hanson, 2004).

Another problem is that cross-sectional measurement necessarily combines results from individuals that may be developing in different ways.

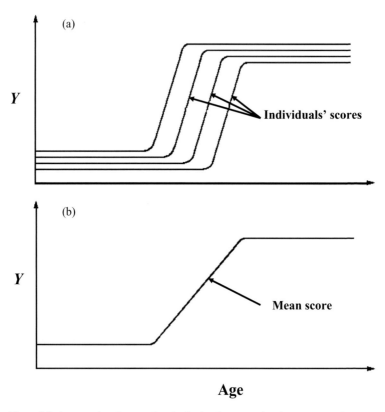

Figure 8.1 A comparison between longitudinal and cross-sectional measurement in developmental studies. Graph (a) shows how a variable (Y) changes as a function of age for four individuals which were measured across time (longitudinal measurement). Each individual exhibits an abrupt increase in Y, but the age at which this starts differs between individuals. Graph (b) shows the mean scores for the same four subjects. These increase gradually as a function of age, a pattern of development that does not represent any individual. A similar picture would have emerged if a different sample of subjects had been measured at each age (cross-sectional measurement). A real example is the sharp increase in the hormone testosterone that occurs in male humans at the time of puberty.

This point is illustrated in Fig. 8.1. Here, a measure (*Y*) increases sharply over a narrow age-range, but the timing of this increase varies between individuals. A cross-sectional study that measured different subjects at each age would show an apparently gradual increase in *Y*, a pattern of development that would not represent any individual. To take a real example, levels of the hormone testosterone increase steeply in human males over a relatively short period (12–18 months) at the time of puberty, but the age at which this increase starts varies considerably between individuals, over a range of several years. A cross-sectional study would give the false impression that testosterone increases gradually over a period of several years.

Both cross-sectional and longitudinal approaches have their advantages and raise different problems of practice and interpretation. Ideally, both methods should be used, as exemplified by the classic study of the development of pecking in domestic chicks by Cruze (1935). He kept the chicks in the dark from when they hatched and fed them by hand. Starting at different ages, he tested the accuracy with which the chicks pecked at seeds. Once a chick had been tested, it was re-tested on subsequent days of its life. In this way Cruze was able to obtain cross-sectional data on chicks that were first tested at a given age, and longitudinal data on chicks that were re-tested each day. Not surprisingly, he found that both a chick's age and its prior experience of pecking at seeds affected its accuracy.

A **sensitive period** in development is an age-range when particular events are especially likely to affect the individual's development. The experimental procedures needed to establish that a given condition is more likely to affect subsequent behaviour at one age rather than at other ages are shown in Fig. 8.2. The age-range when a group of subjects is exposed to the condition is shown by the thick horizontal bars and subsequent testing is denoted by the vertical arrows. If the exposure were started at different ages and ended at the same age, then the age at first exposure would be confounded with duration of exposure. Thus, any observed effect could have arisen because the individuals that were exposed at an earlier age were also exposed for longer. A more subtle difficulty is raised by the time of testing. If the time from the end of exposure to testing is not kept constant, then some of the differences between groups could have arisen because the effects of exposure had more time in which to decline in the groups first exposed at the younger ages. However, if time from exposure to testing is kept constant (as shown in Fig. 8.2), the ages and intervening experiences of the groups at testing are necessarily different. Here again, the counsel of perfection is to use both methods for deciding upon the time of testing.

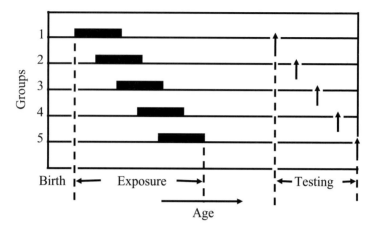

Figure 8.2 An experimental procedure for determining the extent of sensitive periods in development (a stage in development during which an agent or event, such as exposure to a particular type of experience, is especially likely to have an effect). Exposure to the agent or event is denoted by the black rectangles and subsequent testing is denoted by the vertical arrows. In this example, five different groups of subjects have been exposed at different ages. The duration of exposure is the same for all groups. If the time from exposure to testing is kept constant then the age of testing varies. If the age of testing is kept constant, the time from exposure to testing varies.

When experiments are not feasible, development may be investigated by correlational methods. One approach is to study development **retrospectively**, looking backwards in time to what may have been important events in development. In contrast, a **prospective** approach identifies all the individuals that have had a particular experience and examines what subsequently happens to them. These two approaches may yield different results because individuals who were not influenced by the early experience would not be detected by a retrospective examination of those that were influenced.

In many studies of development, differences in behaviour between individuals may be attributed to many things such as genetic or non-genetic inheritance (epigenetic) factors or a host of environmental factors. One technique often used on rodents is to take young born to one mother and have them fostered by another mother, replacing her young by offspring from the second mother. This technique, known as **cross-fostering**, allows you to determine whether the behaviour of the young was influenced by transmission of genetic or epigenetic factors from the biological mother or by the experience of being reared with the foster mother. Of course, the young may be influenced by both.

Tests of preference and differential responsiveness

Many studies of behaviour, both in the field and the laboratory, assess a subject's responsiveness to a range of different stimuli, which might include other individuals of the same species. Such tests of differential responsiveness or preference may be conducted and analysed in a variety of ways, and are known as **choice tests**.

Several issues arise when considering choice tests:

- Should the stimuli be living animals, stuffed animals, audio or visual recordings, or models? Other sensory modalities besides vision can be employed; for example, the stimuli might be smells, tastes or different temperatures.
- Should the stimuli be presented simultaneously or successively?
- Should the same subject be tested once or repeatedly?
- Should the types or rates of response to the stimuli be compared in terms of absolute or relative differences?

Each method has its own advantages and drawbacks and these must be considered carefully before proceeding with a study. We shall consider each of them in turn.

Stimuli

Living animals are the richest and most natural type of choice stimulus, but they have several potential drawbacks. First, the stimulus animals are liable to interact with the test subject, making it difficult to decide who chooses whom. Sometimes these interactions can be eliminated if the stimulus animals are made unaware of the subject's presence by using one-way screens – though a one-way screen may not eliminate non-visual cues such as sounds or odours, which enable the stimulus animals and subject to interact. In one study, for example, a one-way screen was used to examine mate choice in zebra finches (Hunt *et al.*, 1997). The screen allowed the female to see the stimulus male, but the male could not see her, thereby controlling for the effect of female display characteristics on male behaviour. Even if one-way screens are used, the behaviour of the stimulus animals is likely to change with successive tests as they become habituated. If, on the other hand, different individuals are used as models in successive tests, thereby avoiding the problem of pseudo-replication, the individuals will almost certainly behave differently from one another. It should also be remembered that one-way screens act as mirrors and may cause the stimulus animal to act aberrantly. Finally, because a living animal does provide such a rich and complex stimulus, it can be difficult to know precisely which features of the stimulus influenced the subject's choice.

Stuffed animals and recordings have many of the advantages of living stimuli but cannot be influenced by the subject. The effectiveness of particular features of the stimuli can be analysed by selective removal or alteration of different parts of the choice stimuli. This type of manipulation is particularly easy with audio and video recordings. However, the responses of the test animal might be reduced if presented with a stuffed stimulus animal rather than a real one. Video playback has the advantage over models since it is more likely to reproduce the complex motor patterns often involved in threat or courtship displays. In one study, for example, digital video sequences of the complex territorial defence display of lizards successfully replicated the display of the living animals (Ord *et al.*, 2002). Robotic or computer-generated animals can provide a further level of control and can be programmed to interact with the subjects.

Playback techniques have been used with a range of animals, including birds, monkeys and lions, to determine the function of different vocalisations and also to assess more subtle aspects of visual and vocal communication. For example, playing recordings of female Barbary macaque vocalisations to males revealed that male macaques can discriminate between copulatory calls emitted during different stages of oestrus (Semple & McComb, 2000).

Olfactory stimuli have also been used in choice tests. For instance, one study found that female sticklebacks will make subtle discriminations between water-borne scents from different male sticklebacks, spending more time in the part of the tank containing the scent of males whose genetic make-up complemented their own in the major histocompatibility complex (MHC) (Milinski *et al.*, 2005). Diversity in these genes confers resistance against parasites and pathogens.

When choosing stimuli for choice tests you need to be aware of the sensory capabilities of the test species (see D'Eath, 1998). To give just one example, natural variation in ultraviolet reflectance is important in mate assessment by birds (Bennett *et al.*, 1997). Standard laboratory illumination is weak in the ultraviolet wavelengths in comparison with natural daylight. It also flickers – undetectably to humans but detectably by, for example, birds. Similarly the flicker of a cathode ray tube can be detected by European starlings and induce muscle twitches with possible consequences for their welfare (Smith *et al.*, 2005). Using LCD screens is much more satisfactory.

Simultaneous or successive tests?

Simultaneous presentation of test stimuli may be distracting to the subject. Furthermore the subject may become 'trapped' by its first choice, simply because it happens to be facing one way and approaches the stimulus it is facing. Choices may also be affected by strong position preferences. Withdrawing from one stimulus may be incorrectly interpreted as approaching the other (or vice versa). This last problem can sometimes be overcome by providing a three-way choice, with a 'blank' (no-stimulus) choice in addition to the two choice stimuli. Finally, as is well known by salesmen, a human preference can change when a third, less preferred option is added to a binary choice; the same is true for other animals (Bateson, M., 2004). The third option can change the relative preference for the other two or even cause the absolute preference to be reversed.

On the other side of the argument, successive presentation of test stimuli may prove to be insensitive to preferences if the subject responds at the maximum possible level to each of several stimuli when it cannot make simultaneous comparisons between them. More seriously, successive tests may be confounded by order effects because the subject becomes generally more (or less) responsive to all stimuli during the course of testing. Such effects may be allowed for by varying the order of presentation for different subjects, or by presenting two stimuli (A and B) in the order A, B, B, A. When more than two stimuli are used, a randomised Latin Square experimental design can be used to eliminate order-of-presentation effects (Mead, 1988; Ruxton & Colegrave, 2003). Further discussion of choice tests is given in Chapter 10.

Composite measures

It can sometimes be informative to combine two or more mutually exclusive behavioural measures that are thought to be alternative expressions of a single, underlying propensity. For example, when presented with a threatening stimulus an animal may react in a number of different and mutually exclusive ways, such as attacking, freezing or fleeing. If good grounds exist for supposing that any one of these alternative responses indicates the same underlying motivational state (fear), then pooling these measures would make biological sense, even though they do not overtly measure the same behaviour patterns and are not positively correlated with one another.

If the benefits of combining measures are likely to outweigh the disadvantages, such as discarding potentially helpful information, how should the component measures be chosen and, once chosen, how should they be combined? Picking the component measures is often done intuitively or on the basis of other knowledge. A more systematic approach is to construct a matrix to see which measures tend to be inter-correlated and then apply multivariate statistical techniques (see Chapter 9). The measures that are to be combined usually need to be standardised so that they have the same mean and variation. One way is to calculate for each raw value its z score: the score for that individual minus the mean score for the sample, divided by the standard deviation. Scores standardised in this way have a mean of zero and a standard deviation of 1.0. The composite score for an individual is then the average of the z scores of the separate measures. This procedure gives the same statistical weight to each measure, but different weights could be applied if such a procedure could be properly justified.

How much information to collect?

Up to a point, the more data collected the better, because statistical power is improved by increasing the sample size. However, increasing the sample size by amassing more and more results eventually offers diminishing returns in terms of increased statistical power: when sufficient results have been acquired, additional results may add little to the ability to draw sound conclusions. The temptation to carry on collecting results indefinitely must therefore be weighed up against the time and effort involved, since at some stage it will be more productive to move on to a new study than to keep collecting additional data for the current one.

The problem lies in knowing what constitutes sufficient data. Certain simple rules of thumb can be used. In general, if the number of degrees of freedom (defined in Appendix 2) is less than 10, the sample sizes will probably be too small to generate a reliable result, whereas if it is greater than 20, little useful information will be added by increasing the sample size.

Some authorities that oversee the use of animals in research and some funding bodies may demand what is called a **power analysis** before research can proceed or be funded. The power analysis procedure allows you to estimate the probable sample size required for an experiment. For instance, if two groups are to be compared you need to provide values for the means of both groups and the pooled standard deviation to feed into a formula. Apart from requiring knowledge that may not be available in advance of conducting the research, such a procedure is often inappropriate in the case of behavioural work because its underlying assumptions are violated. If you are stuck, we suggest you use the rule of thumb described above. Alternatively, read an authority on power analysis such as Bausell and Li (2002) and, despite the reservations, apply the recommended formulae to satisfy those who demand it.

Summary

All research benefits from careful design. Even in the simplest experiments, thought should be given to control groups and the use of blocks in designs so that the most plausible alternative hypotheses to the one being tested can be eliminated. Factorial experiments in which several treatments are varied simultaneously require fewer subjects overall than when the treatments are varied separately; they may also reveal interactions between the different treatments. Matching subjects in pairs or repeated testing of the same subjects have some benefits, but both designs carry drawbacks that should be considered in advance. Studying behavioural development requires special techniques because the subjects vary over time as they mature. Choice tests require thought to be given to the character of the stimuli and whether or not they are presented simultaneously or successively. In all studies in which several dependent variables are measured, composite measures may sometimes be constructed. Estimates of how much information to collect can sometimes be obtained in advance but they depend on assumptions that may not be justified. A simple rule is that when the degrees of freedom exceed 20, the benefits of increasing the sample size will diminish.

9

Statistical analysis

General advice on statistics

We appreciate that most people who read this book will learn most about statistics when they start to analyse their own data. In general, we advise against poorly informed cook book approaches to statistical analysis. Our aim in this chapter is to consider briefly some of the main issues that arise in the statistical analysis of behavioural data. This is not a statistics textbook, however, and for an account of statistical methodologies you should consult one of the many excellent books available, but preferably one that has as its target a biological or psychological audience (e.g. Zar, 1999; Sprinthall, 2003). In Appendix 3 we have given an annotated list of books that we have encountered.

Given the inherent variability in biological systems, statistical analysis is often essential for unravelling what is going on. Nonetheless, excessively complicated statistics are sometimes used as a substitute for clarity of thought or good research design. Do not become obsessed by statistical techniques, nor too cavalier in their use. Statistical analysis, no matter how arcane or exquisite, can never replace real data.

Besides sometimes being *over*-used, statistical techniques are frequently *mis*used in the behavioural literature. We have already outlined one common error in Chapter 7 – that of including many data points from the same individual in the mistaken belief that they are independent measurements. In Chapter 11 we consider, among other things, the misinterpretation of multivariate statistics and the various misuses of correlation coefficients.

The choice of behavioural categories and recording methods to be used in a study should take account of the statistical methods that will eventually be used to analyse the data. Sometimes results are painstakingly collected, only to be found unsuitable in form for the statistical tests required to analyse them. As already noted in Chapter 8, consulting a statistician (preferably one who is familiar with behavioural research) *before* starting to collect data is generally a wise precaution. Of course, statisticians do not always agree with each other. Occasionally they insist on inappropriate methodologies for a particular area of research, or they focus on details that do not affect the overall outcome of the analysis. Nevertheless, our advice stands.

Spreadsheets and databases

Once observations or experiments have been completed, the next step is to convert them into a form in which the data may be readily examined. We cannot over-emphasise just how important it is to consider this phase of your work carefully. If you do it well, you can save yourself a great deal of time at a later stage.

Spreadsheets are convenient for bringing all your data together. You should aim to include information that may be required to make corrections at a later stage. So, for example, if you have conducted a field study in which you have watched different individuals for different amounts of time, you should include a column in the spreadsheet specifying how much time you watched each individual. Subsequently the number of times you have witnessed a particular form of behaviour expressed by an individual can be corrected for the total time spent observing it. A spreadsheet for a simple experiment might have a column for identifying each subject, a column for the type of treatment (e.g. control or experimental), a column for the sex of each individual, a column for time tested, a column for the date of the test and separate columns for each behavioural measure. Each cell in the measurement columns will then contain the data from one subject for one measure. Remember to enter data about each subject such as its sex and any other categorised or quantified information that may be used in subsequent analysis. The principle here is the opposite of packing a suitcase: if in doubt put it in. More complex experimental designs can be accommodated by entering columns for each block or for each combination of factors. Spreadsheets allow data to be examined visually and subjected to simple exploratory statistical analysis (see below). Data can also be readily pasted into statistics packages such as Minitab, SPSS or SAS.

Other software can allow the creation of flexible databases that communicate with the spreadsheets into which data have already been entered. Relational database programs such as Access allow information to be stored in multiple files or tables. Virtually any combination of data can be extracted from the database, with many sorting options. The combination may depend on multiple selection criteria, avoiding the frequent pasting required if a spreadsheet alone is used.

Exploratory versus confirmatory analysis

Statistical techniques are used for two quite different purposes: **exploratory** data analysis, and **confirmatory** data analysis. Most conventional statistics textbooks deal mainly with confirmatory analysis.

Exploratory data analysis (or descriptive statistics) includes the essential – but often neglected – processes of collating, summarising and presenting results, and searching through them to extract the maximum amount of information. This is especially important when the results are complex or the hypotheses vague. Exploratory analysis provides a way of learning from results and generating new questions or hypotheses from them. The severe advice given in many textbooks is that hypotheses should not be generated and then tested using the *same* data, a procedure that one distinguished statistician described as tantamount to offering to bet on a horse race after the finish of the race. While the point is well made, it often amounts to a counsel of perfection. A data set may have been explored even though the same data are used to test hypotheses formulated in advance of collecting the data. Furthermore some confirmatory procedures (described below) involve exploring possible statistical models that best describe the data.

One of the simplest and most fruitful types of exploratory analysis is plotting results in the form of a graph, histogram or scatter plot. It is always wise to plot results and inspect them visually before plunging into confirmatory statistical analysis. As a general rule, graphical or other visual summaries are more informative than dense tables of figures. It is also helpful to summarise data in the form of descriptive statistics, such as means or medians and standard deviations or ranges, before hypothesis-testing is carried out.

Confirmatory data analysis (hypothesis-testing, or inferential statistics) covers the conventional 'testing' of empirical data – that is, calculating the probability that the observed result is consistent with a null hypothesis such as 'there is no difference between the mean scores of the two groups', or 'there is no association between the two sets of scores'. If this probability is lower than the pre-determined level (usually 0.05) then the null hypothesis is rejected.

The main purpose of hypothesis-testing is to provide a publicly understood way of specifying how much confidence can be placed in an apparent effect such as a difference or a correlation. Confirmatory data analysis is central to studies of natural variation as well as experiments, since it includes tests of hypotheses about correlation.

Unless only small quantities of data are involved, confirmatory data analysis is usually performed using one of the standard statistical software packages. However, we urge caution. Just because it is easy to do so, data should not simply be fed into a statistical package and analysed in every possible way without careful thought about the questions being asked. Moreover, you should become familiar with the data through exploratory analysis before subjecting them to confirmatory analysis in what, for many, will be a black box provided by the software. Some traditionalists would urge you to plot out the data by hand and even do some elementary calculations to provide yourself with a good 'feel' for what you have discovered.

Data analysis is not a purely mechanical exercise. A set of data can always be analysed in many different ways: and as findings are compared with hypotheses and expectations, new ideas arise which stimulate the analysis of the data in other ways. If the data are rich and complex, the process of analysis and interpretation may be a gradual one. It is undoubtedly a very important phase.

What statistical tests should be used?

Two basic classes of tests are available – parametric and non-parametric. The parametric tests depend on a number of assumptions that are often violated in behavioural research. Non-parametric tests have often been preferred in the past because they are not so demanding of the data and are therefore more robust and realistic. The most commonly used source for advice on non-parametric tests is Siegel and Castellan (1988). Despite the past popularity of non-parametric statistics among behavioural biologists, parametric tests have been used much more frequently in behavioural work in recent years as scientists have used more efficient experimental designs or faced more complex arrays of data to analyse.

Unlike parametric tests, which calculate exact numerical differences between scores, non-parametric tests consider whether particular scores are higher or lower than other scores. Since non-parametric tests require only ranks rather than measurements on an interval or ratio scale, they can be used to analyse data measured on an ordinal scale (see Chapter 3). Some tests, such as chi-square, can be used to analyse data measured on a nominal scale. For simple tests of whether two groups differ from each other or whether two variables are associated, non-parametric tests will continue to be used. The differences from parametric tests in their power to detect statistically significant effects are often marginal. In any case, if a non-parametric test is powerful enough to detect a significant effect, then it is powerful enough – or, as one statistician put it: 'If the tree falls, the axe was sharp enough.' In addition to their relative freedom from unrealistic assumptions, non-parametric methods have the advantage over the equivalent parametric methods that they are better suited to working with small ($n < 10$) sample sizes – a situation that behavioural scientists regrettably face all too often. Parametric statistical analyses have to be used for more complex types of analysis than those that look simply for differences or associations.

Parametric tests are generally based on the following assumptions about the nature of the population from which the sample data are drawn:

- *Normality*. The data follow a normal (Gaussian) distribution.
- *Homogeneity of variance*. Different samples or sub-groups vary to the same extent.
- *Scale of measurement*. The data are measured on an interval or ratio scale.
- *Additivity*. The effects of different treatments or conditions are additive (see below) and associations between independent and dependent variables are linear. This assumption is not made in the case of General Linear Models (GLM).

Behavioural data frequently violate some or all of these assumptions to some extent, thereby calling into question the validity of parametric methods. For example, behavioural data sometimes have a highly skewed distribution, violating the assumption that data are normally distributed. In practice, however, the consequences of moderate departures from a normal distribution may not be serious. Furthermore, data can be made to follow a more nearly normal distribution by applying a suitable transformation. You should certainly test your data to see whether such transformation is necessary. Procedures for doing so are described in many text books.

The guidelines for transforming data are as follows:

Square root transformation. Data in the form of *counts* – for example, true frequencies or total numbers of occurrences – are likely to follow a Poisson rather than normal distribution. This type of non-normality can often be corrected using the square root transformation – i.e. convert x to $\sqrt{x}$. If the data include zero scores, then add 0.5 to all values – i.e. convert x to $\sqrt{(x + 0.5)}$.

Arcsine-square root transformation. When data are in the form of *proportions* or *percentages* (for example, when time sampling is used) non-normality can often be cured by transforming data according to the arcsine-square root (or angular) transformation – i.e. convert each value to arcsine($\sqrt{p}$), where p is a proportion ($0 < p < 1$). If the scores fall between 0.3 and 0.7, then it is not usually necessary to use the transformation.

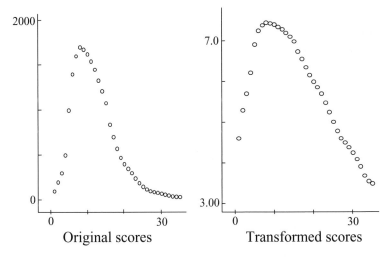

Figure 9.1 A distribution of data before and after they have been transformed by taking, in this case, natural logarithms. The original scores have more of a skew than the transformed scores, which approximate to a normal (Gaussian) distribution.

Logarithmic transformation. The logarithmic transformation – i.e. convert x to $\log(x)$ – is useful in a number of cases; for example, when the mean and variance are positively correlated or when the frequency distribution is skewed to the right. Commonly, natural logarithmic transformations are used rather than logarithms to the base ten. If the data include zero scores, add 1 to each score – i.e. convert x to $\log(x + 1)$. An example of a transformation is given in Fig. 9.1.

None of the common parametric tests depends solely on the normality of data, so normalising data does not by itself ensure their validity. However, transforming data will sometimes cure more than one type of departure from the assumptions of parametric tests, such as making data fit the assumptions of homogeneity of variance as well as normality. Indeed, the primary benefit of many transformations is that the variances of different groups or sub-groups are made more similar to each other.

Analysis of variance (ANOVA)

Analysis of variance is needed when analysing the effect of an independent variable on a dependent variable (see Chapter 8). At its simplest it can be used to compare the difference between two groups, and it yields exactly the same result as a Student's t test. When many comparisons are possible, ANOVA is greatly preferred to making many independent tests, because the probability of one of the tests being significant by chance alone becomes unacceptably high (i.e. > 5%). Using ANOVA thereby allows comparisons to be made between any number of groups in a single test. If appropriate corrections are made, differences between two particular groups can subsequently be tested. A commonly used method of correction is the **Bonferroni test** in which all pair-wise comparisons are examined. The specified level at which the null hypothesis is to be rejected is divided by the number of possible comparisons. For example, if the level of probability below which a result is to be regarded as statistically significant is 0.05, and the number of pair-wise comparisons is three, then the corrected level is $0.05/3 = 0.017$. By convention, anything above this would not be regarded as statistically significant. Consider a case where a measure of maternal behaviour in three different groups had been taken at three different times after birth. Suppose the probability that a difference between the group measured at one day after birth and the group measured 14 days after birth because of chance arose with a probability of 0.03 in a pair-wise comparison without correction. This would not normally be regarded as statistically significant.

When the independent variable is only measured on a nominal scale, such as sex or place, ANOVA is always used. It often provides a better approach in cases where the assumption of a linear relationship between independent and dependent variables is inappropriate. In the example of maternal behaviour given above, such non-linearity would be the case if the score rose from day one to day seven after birth and then fell again.

Correlation

A correlation coefficient describes the extent to which two measures (or variables) are associated or vary together. Two measures – for instance, height and weight – are positively correlated if high scores on one measure are associated with high scores on the other measure, and low scores on the first are associated with low scores on the second. Conversely, if high scores on one measure are associated with low scores on the second and vice versa then the two measures are negatively correlated.

The strength of the association is indicated by the size of the correlation coefficient, which is a number between -1.0 and $+1.0$. A correlation of ± 1.0 indicates a perfect association – i.e. every score on one measure is perfectly predicted by the scores on the other measure. A correlation of 0 means there is no linear association between the two measures – i.e. knowing one set of scores provides no predictive information about the other set. A *significant* correlation is generally taken to mean a correlation that differs significantly from zero. The distinction between statistical significance and biological importance is discussed in Chapter 11.

For an illustration of the strength of association represented by correlations of different sizes, see Fig. 9.2. Note that at first sight a weak but statistically significant correlation (Fig. 9.2*d*) may appear little different from a complete absence of correlation (Fig. 9.2*c*).

Correlation coefficients should not be averaged

Correlation coefficients are not like ordinary numbers and they do not obey the normal rules of arithmetic. In particular, it is incorrect to average several correlation coefficients by calculating their arithmetic mean. However, the *difference* between two parametric Pearson correlations is meaningful, and its statistical significance can be tested.

Correlation coefficients cannot be directly compared

Two correlations cannot be directly compared as a ratio in the simple way that, say, two weights or two distances can be compared. For example, a correlation of 0.8 does not represent an association that is twice as strong as a correlation of 0.4. One valid way of comparing Pearson correlations is by using the square of the correlation coefficient (r^2), which is known as the **coefficient of determination**. Broadly speaking, r^2 is the proportion of the variation in one measure that is accounted for statistically by the variation in the other measure. Hence, a correlation of 0.8 means that 64% of the variation in one set of scores is accounted for statistically by the variation in the other ($r^2 = 0.64$). With a correlation of 0.4, one measure

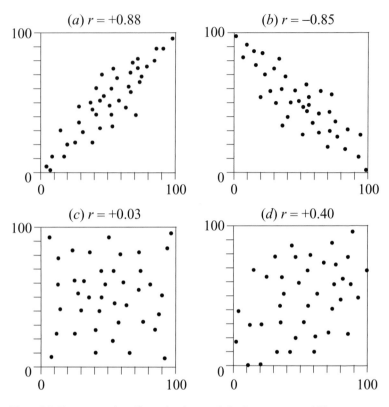

Figure 9.2 Four scatter plots, illustrating the association between two variables represented by various correlations. All correlations are Pearson coefficients (r) with 38 degrees of freedom ($n = 40$). (a) $r = +0.88$ ($p < 0.001$), (b) $r = -0.85$ ($p < 0.001$; (c) $r = +0.03$ (not statistically significant; $p > 0.10$); (d) $r = +0.40$ ($p < 0.02$). Note that although the correlation shown in (d) represents a weak association, it is, nonetheless, statistically significant.

accounts for only 16% of the variation in the other measure ($r^2 = 0.16$). In this sense, a correlation of 0.8 is *four* times as great as a correlation of 0.4.

Simple regression

Simple regression provides a powerful technique for describing and analysing the relationship between an independent variable (X) and a dependent variable (*Y*). Its main value lies in enabling the value of the independent variable to be predicted for a given value of the dependent variable. In regression analysis, which is sometimes confused with correlation, the dependent variable is expressed as a mathematical function of the independent variable: $Y = f(X)$.

In its most basic form, **simple linear regression**, the relationship between *Y* and *X* is a straight line. The **regression line** is the best possible fit of a straight line relating the two variables – that is, the straight line which lies closest to all the points on a scatter plot of *Y* against *X*. This line is described by a linear regression equation:

$$Y = bX + a$$

The slope of the regression line, b, in the equation above, otherwise known as the **regression coefficient**, is the rate at which the dependent variable changes in relation to the independent variable: i.e. the average change in *Y* for a unit change in *X*. The intercept of the regression line, a, represents the predicted value of *Y* when *X* is zero. Having calculated this equation (known as 'the regression of *Y* on *X*') from a set of empirical data, it is a straightforward matter to predict the value of *Y* for any given value of *X*. The higher the correlation between *X* and *Y*, the more closely the points on the scatter plot of *Y* against *X* will cluster around the regression line and the more accurate this prediction will be. In addition to the basic assumptions of normality and linearity, the variance in *Y* scores must be roughly constant for all values of *X* (a property known as **homoscedasticity**).

A straight-line relationship between the dependent and independent variables is the only one tested when a simple linear regression model is applied. In many cases, however, the independent and dependent variables are manifestly not related in a linear manner. In cases like this, where the relationship between the two variables is non-linear, it may be possible to apply a **curvilinear regression**. Instead of expressing the dependent variable as a linear function of the independent variable, it is expressed as a polynomial function of the general form:

$$Y = a + bX + cX^2 + dX^3 + \ldots$$

For example, a parabolic relationship between two variables can be described by a quadratic regression ($Y = a + bX + cX^2$). As higher-order powers of X are added, the regression curve becomes increasingly complex and should fit the data increasingly well.

The process of fitting a curvilinear regression equation to a set of data, known as **curve-fitting**, is an empirical procedure. Starting with a simple linear regression, quadratic (X^2) and, if necessary, cubic (X^3) and higher-order terms, are added in a stepwise fashion until a sufficiently close fit is obtained. Fitting anything much beyond a quadratic regression is usually best avoided unless such relationships make some theoretical sense. An alternative to curve-fitting is to **transform** the data for one or both variables such that their relationship becomes linear – for example, by converting values of Y to $\log(Y)$, $\sqrt{Y}$, $1/Y$ or arcsine (Y). A suitable transformation can often be found that will convert a non-linear relationship into a linear one, allowing a simple linear regression model to be applied.

General linear models (GLMs)

The trend among behavioural biologists in recent years has been to switch to the powerful general linear models (GLMs). These statistical models are described clearly in Grafen and Hails (2002). Their big advantage is that they combine the techniques of correlation, regression and ANOVA described above, they do not require that all the groups in an experiment contain exactly the same number of individuals and, above all, they allow you to explore the statistical model that best fits your data. This is particularly important when you are investigating interactions between independent variables. You might have found, for example, that males score lower on average than females at the lower end of one set of conditions and higher than females at the higher end of the same set of conditions. The GLM enables you to develop the appropriate statistical model that demonstrates this interaction. Some statistical approaches assume that the effects of the independent variables on the dependent variable are additive. The GLM does not because it allows you to find a model that captures non-additive effects. As Grafen and Hails (2002) emphasise, one of the main points of the GLM approach is to have a conceptually powerful and easily implementable way of testing assumptions of additivity.

Multivariate statistics

In most studies of behaviour, several independent as well as dependent variables are used. The statistical techniques used for analysing data where more than two variables are involved are known collectively as **multivariate statistics**. They can generally be implemented using the GLM approach described above. Although this is an enormous topic in its own right, and far beyond the scope of this introductory text, some brief comments are in order.

By analogy with simple linear regression, described earlier, the technique of **multiple regression** is used when a dependent variable (Y) is thought to be related to two or more independent variables ($X_1, X_2, \ldots, X_n$). For example, the height of a child might be thought to depend on both the height of the mother and the height of the father. Multiple regression allows Y to be predicted from measures of two or more independent variables, by regressing Y simultaneously against all of them. This is expressed in a multiple regression equation:

$$Y = a + b_1 X_1 + b_2 X_2 + \ldots + b_n X_n$$

The coefficients (b_1, b_2, etc) are known as **partial regression coefficients**. For example, b_1 represents the regression coefficient of Y on X_1 that would be expected if all the other independent variables had been held constant.

Multiple regression techniques can be extremely helpful in deducing causal links, particularly when experiments have not been possible. **Path analysis** is an extension of the regression model, used to test the fit of the correlation matrix against two or more causal models which you might wish to compare. Here again, the plausibility of the answers obtained with such techniques depends on whether or not the assumptions of the test are valid.

By analogy with simple correlation, a **multiple correlation coefficient** expresses the degree of covariation between three or more inter-related variables. Similarly, **partial correlation** can disentangle the mutual dependence of a set of variables by showing how any two are related when the effects of the others have been eliminated. A partial correlation coefficient is a measure of the correlation between any two variables when the others are kept constant – for example, the correlation between Y_2 and Y_3 when Y_1, Y_4, etc. are constant. Partial correlation can be used to identify spurious associations that arise when two variables are both correlated with a third. Two variables that are significantly correlated with each other may be shown to have a non-significant partial correlation when a third variable is held constant. In such a case, it is likely (but not certain) that the third variable is a common cause underlying the original correlation. So, for example, the foraging rate of a bird might be found to be correlated with both the time of day and the ambient temperature, but when time of day is held constant, temperature might be found to be unimportant.

Correlation matrices can be constructed if measures have been obtained for three or more variables. This matrix consists of a table of the correlation coefficients of each variable with every other variable. This might be done in order to uncover significant correlations that had not been predicted (but see Chapter 11), and provides a route to exploring relations between the variables. In many cases the variables will be linked noticeably together in clusters. Various techniques have been developed to describe and analyse these groupings according to the magnitudes and inter-relationships among their correlation coefficients.

Factor analysis methods, of which one main type is **principal component analysis (PCA)**, provide one way of uncovering groupings. They aim to reduce the complex inter-relationships between a large number of variables down to a smaller number of underlying factors that account for a large proportion of the variance and covariance of the original variables. In this respect PCA identifies which combinations of variables explain the greatest amount of variation in the multivariate data. The PCA method usually treats each variable as equally important. New variables or principal components (PC) are linear combinations of the original ones, and are created to explain the largest possible amount of information in the data. The first PC explains the largest amount of information; the second

PC explains the second largest, while being as different as possible from the first PC; and so on. The values of the principal components are plotted against each other as a scattergram, thereby enabling differences between the types of data to be assessed.

Other techniques such as **cluster analysis**, which helps identify patterns within the data, and **discriminant analysis**, which classifies variables into groups, are best understood by consulting an advanced textbook.

Independent variables may consist of both categorical data and continuously distributed data. A method that takes account of both these is known as **analysis of covariance** (ANCOVA) and may be used to clarify relationships between independent and dependent variables. When two variables influence a third, their separate influences on the outcome may obscure each other. Analysis of covariance can remove the effect of one variable and, in so doing, make the effect of the second more obvious. Such statistical methods are particularly helpful when the contribution of a 'nuisance' factor, which also affects the dependent variable, needs to be minimised before the analysis of variance takes place. If, say, the foraging rate of a bird was affected by age and the time of day, ANCOVA would allow the age effect to be removed in order to discover whether that strengthened the effect of time of day.

By analogy with ordinary analysis of variance (ANOVA), **multivariate analysis of variance (MANOVA)** is used where a number of independent variables have been measured for each of several samples. It is a technique that performs analysis of variance on more than one dependent variable and explicitly takes into account the correlations between the dependent variables. By considering all scores simultaneously, MANOVA may be able to detect a subtle difference between groups, even when none of the univariate analyses does so. The multivariate test helps to decide whether such a difference is caused by different relationships among the dependent variables, or just one underlying mechanism that is being measured in several different ways.

Finally, it is worth emphasising that for multivariate analysis the number of independent subjects should be relatively large in comparison with

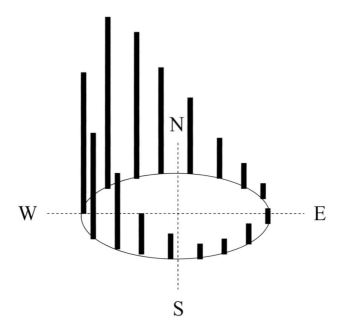

Figure 9.3 Distribution of departure directions of animals moving away from a central point in this hypothetical case. The greatest number of animals, indicated by the height of the black bars, moved in a north-westerly direction. The two tails of the distribution meet at the south-easterly compass point, which is why the number moving in that direction is slightly higher than on either side.

the number of independent variables. Unfortunately, for practical reasons, behavioural scientists sometimes produce data that refer to relatively few subjects and a large number of variables.

Circular statistics

Some studies collect data in which the two tails of the distribution meet because the distribution is circular. For example, departure paths from a given position could be in any compass direction. If they are normally distributed around a particular direction, the two tails of a symmetrical distribution could in theory meet at a direction which is 180° away from the mean (see Fig. 9.3). Circular statistics are designed to deal with this problem (Jammalamadaka & SenGupta, 2001) and are especially useful in analysing data from studies of orientation movements.

Did you collect enough data?

When planning a study, you should have some idea of how many subjects will be required and, as we indicated in Chapter 8, some rules of thumb for making such estimates. However, you may be unable to tell in advance how variable the subjects are and, therefore, how many will be required to decide whether or not a particular condition affected them.

A simple way of checking whether sufficient data have been collected is to divide the data randomly into two halves and analyse each half separately. If the two sets of data both generate clear conclusions that are in agreement, then sufficient results have probably been obtained. If, however, the two sets of data lead to conflicting conclusions, or are insufficient to produce any firm conclusions, then more data are almost certainly needed. A more sophisticated approach, known as **split-half analysis**, is to divide the data for a particular category of behaviour randomly into two halves and calculate the correlation between the two sets of data. If the correlation coefficient is sufficiently high (say, $r > 0.7$), then the data set is said to be reliable.

If a study produces a result that borders on statistical significance (e.g. $0.05 < p < 0.10$), then it may be wise to increase the sample size in order to reach a definite conclusion about whether the effect is real or not. This should be done by deciding *in advance* by how much the sample size should be increased. The practice of making piecemeal increases in the sample size until the results 'reach significance' is dubious, since it increases the chances of obtaining a spuriously positive effect. When a study fails to reveal any significant effects, post-hoc power analysis can be conducted to help find out whether adequate sample sizes had been used (Bausell & Li, 2002). However, as we noted in Chapter 8, the assumptions of such analysis are frequently violated by behavioural data.

Summary

If information obtained from a study is initially entered into a spreadsheet, subsequent work is made much easier. Statistical techniques are invaluable tools but, in general, you would be wise to keep them as simple as is practicable. A distinction can be drawn between exploratory analysis and confirmatory tests that are used to test a specific hypothesis. While dependent variables that are nominal or ranked must be tested with non-parametric tests that make few assumptions, parametric techniques have developed so much that they are now commonly used in behavioural studies. Their use may require that the data be transformed so that assumptions of the tests are not violated. Analysis of variance, correlation and simple regression are commonly used. The general linear models (GLMs) incorporate all these techniques and by iterative steps allow the flexible development of the simplest and most appropriate statistical model that takes account of interactions between the independent variables. Multivariate and circular statistics allow complicated analyses of more complex or unusually distributed data sets. Simple procedures can be used for checking whether enough data have been collected for a result to be reliable.

10

Analysing specific aspects of behaviour

In this chapter we consider some particular aspects of the analysis of behavioural data. We begin with the crucial dimension of time. We consider a number of ways in which order can be extracted from the observed stream of behaviour. We go on to consider how best to treat the data obtained from choice tests as described in Chapter 8 and conclude with some ways of dealing with social behaviour.

Bout length

An estimate of bout length may be required when behavioural acts recur in temporal clusters (a bout of *events*) or when the same, relatively prolonged behaviour pattern occurs continuously for a period (a bout of a single behavioural *state*). If behaviour patterns are neatly clumped into discrete bouts separated by uninterrupted gaps, then one bout can be distinguished from the next with relative ease. Often, though, bouts are not obviously discrete, in which case a statistical criterion must be used to define a single bout of behaviour. One commonly used technique is **log survivorship analysis**. This is a simple graphical method for specifying objectively the minimum interval separating successive bouts: the **bout criterion interval (BCI)**. Any gap between successive occurrences of the behaviour that is less than the BCI in length is treated as a *within-bout* interval, while all gaps greater than the BCI are treated as *between-bout* intervals. To estimate the BCI, the cumulative frequency of gap lengths (on a logarithmic scale) is plotted against gap length (on a linear scale). Fig. 10.1 shows a stylised example of a log survivorship plot.

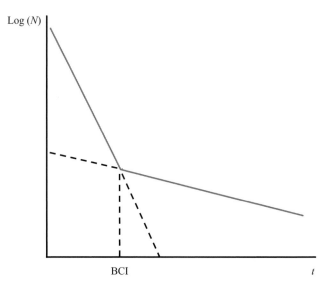

Figure 10.1 The general form of a log survivorship plot, used for determining bout length. The time interval between successive occurrences of the behaviour pattern is *t*. Log (*N*) is the logarithm of the number of intervals greater than the corresponding value of *t*. The bout criterion interval (BCI) is an objective estimate of the maximum within-bout interval and the minimum interval that distinguishes separate bouts.

If the assumptions of the statistical model are met, the log survivorship graph should have two fairly distinct parts: a rapidly declining portion representing the short within-bout gaps, and a slowly declining portion representing the long between-bout gaps. The best criterion is an interval that corresponds approximately to the break point. If the slope changes gradually and no clear break point occurs on the graph, then splitting the behaviour into bouts is probably wrong.

Analysing sequences

For some purposes, the issue of interest is the sequence of behaviour, not the frequencies or durations of the component behaviour patterns (Bakeman & Quera, 1997). An investigator may wish, for example, to record the sequence of display movements in the courtship behaviour of a bird. If the various behaviour patterns in a sequence are mutually exclusive – that is, only one can occur at any one time – then recording the sequence simply involves noting down each occurrence. With only three different behaviour patterns (*A, B* and *C*), the record might be of the form: *ABAAACCBBAAACA* . . . , in which no obvious order is apparent.

If time-sampling methods are used, information about the sequence of events within each sample interval is generally lost, although some sequence information is preserved on a crude time scale, accurate to the nearest sample interval. We do not recommend time sampling for recording sequences since it is possible to miss some types of behaviour altogether using these methods, unless very short sample intervals are used (see Chapter 5).

A sequence in which the component behaviour patterns always occur in the same order is described as **deterministic**. In reality, sequences of behaviour are usually variable, but exhibit some degree of predictability: these are referred to as **stochastic** or **probabilistic** sequences. Sequences that show no temporal structure, where the component behaviour patterns are sequentially independent of one another, are referred to as **random** sequences. In a random sequence, one behaviour pattern can be followed by any other (including itself) with equal probability. The conditional probability that one behaviour pattern follows another is referred to as a **transition probability**.

Markov analysis is a method for distinguishing whether a sequence of behaviour is random or contains some degree of temporal order. A first-order Markov process is one in which the probability of occurrence for the next event depends only on the immediately preceding event. If the probability depends on the two preceding events then the process is described as second-order, and so on. Sequences are analysed by comparing the actual number of times each transition occurs with the number of such transitions that would be expected if the sequence were random. To take a highly simplified example, suppose only two different acts (*A* and *B*) can occur. If each occurrence is independent of the previous one, *A* is just as likely to be followed by *B* as it is to be followed by another *A*. In other words, when the sequence is random the probability that *A* will be followed by *B* is 0.5. To test whether a sequence is nonrandom, a **transition matrix** is constructed, showing the actual transition

Sequence: ABABABBABABAABABABABA

1st behaviour pattern

		A	B
2nd	A	0.1 (0.5)	0.9 (0.5)
behaviour pattern	B	0.9 (0.5)	0.1 (0.5)

Figure 10.2 A simplified transition matrix for the sequence of behaviour patterns shown above. The matrix shows the transition probabilities for the four different types of transition ($A|A$, $A|B$, $B|A$, $B|B$). For example, the bottom left-hand cell shows the conditional probability of B given that A has occurred ($B|A$) is 0.9. A will be followed by B nine times out ten. By chance the probability would be 0.5. The analysis confirms that A and B tend to alternate.

probabilities. Suppose, for example, that B frequently follows A and vice versa – i.e. the transition probabilities of AB and BA are high and the transition probabilities of AA and BB are low. A transition matrix of the type shown in Fig. 10.2 might be obtained. A real matrix would generally encompass more than two types of behaviour and would therefore have more than four cells.

If the actual transition probabilities differ significantly from the chance level (in this case, 0.5), then the behaviour patterns are not sequentially independent. A chi-square test is sometimes used to calculate whether, overall, the observed transition probabilities depart significantly from the random model, although the assumptions on which this test is based are often violated by behavioural data.

If a sequence is scored in such a way that the component acts can repeat themselves (as, for example, in the sequence *ABBBAAB* . . .) then problems can arise in deciding whether a given act has stopped and started again, or whether the same act is merely continuing (*BB* versus *B*). The way in which this problem is handled greatly affects the conclusions drawn from a transition matrix. One solution is simply not to consider repetitions, but to deal only with transitions between different behaviour patterns, although this is, of course, only helpful when dealing with three or more behaviour patterns.

Transition matrix analysis assumes that the transition probabilities remain constant across time (the assumption of **stationarity**). However, this assumption is often violated, especially in long sequences of behaviour, or sequences describing interactions between two or more individuals. The internal consistency of sequential data can be tested by separately analysing data from different parts of the record.

Some repeated patterns of acts are difficult to detect, either with the naked eye or by using standard analytical methods. For example, a sequence of acts might consist of *AXBYYCXAZYBXC* . . . where the sequence *ABC* occurs regularly but its components are separated by other acts, *X, Y* and *Z* that do not recur in regular sequence. **Theme** is a relatively new technique devised to cope with this and extract and visualise the regularities (Magnusson, 2000). Noldus Information Technology provide software to find such regularities (www.noldus.com).

Analysing rhythms

Rhythmic variations in behaviour and in physiological processes are found in many species, with periods ranging from a few minutes to several years. The most familiar type of behavioural rhythm is the **circadian rhythm**, with a period of approximately 24 hours (see Fig 10.3). However, rhythmic variations over different time scales are also found, including **ultradian rhythms** (periods considerably less than 24 hours), **infradian rhythms** (periods considerably more than 24 hours, such as lunar cycles) and **circannual rhythms** (periods of approximately one year, such as annual migration or hibernation).

A behaviour pattern is said to be **rhythmic** if it is repeated over time such that the distribution of intervals between successive occurrences is roughly regular, or if the rate of occurrence varies in a roughly cyclical manner. If the intervals between successive occurrences are approximately equal, within specified limits of variation, then the behaviour is

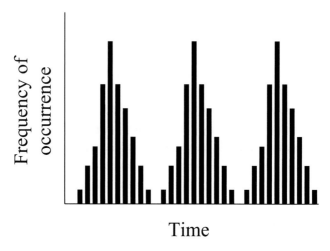

Figure 10.3 A hypothetical example of rhythmic behaviour. Each bar refers to the number of times the behaviour was observed in a given block of time.

said to be **periodic** (or cyclic). The term periodic is more restrictive than rhythmic and is applied only when the regularity of variation has been demonstrated. A simple example of a periodic function is a sine wave.

Periodic variations over time are conventionally described using the same terms that are applied to vibrations or waves. The **period** (or wavelength) of a rhythm is the interval between successive peaks (or troughs) on the wave. The **amplitude** is the magnitude of change between peaks and troughs: that is, the maximum range of variation in behaviour. Many behaviour patterns exhibit complex rather than simple rhythms, with several rhythms of different periods superimposed on one another. These correspond to the harmonics superimposed on the fundamental oscillation of a vibrating string. One or more ultradian rhythms in activity are often superimposed on a dominant circadian rhythm.

Rhythmic variations in behaviour can be detected using four principal methods of analysis:

Plotting the data is the simplest method of looking for rhythms and involves plotting the behavioural variable as a function of time. A periodicity of large amplitude should, if present, be fairly obvious.

Autocorrelation involves calculating all possible correlations between values of the variable at different intervals and is an effective way of detecting the dominant rhythm in a time series. If a variable oscillates with period T, a maximum positive correlation between the successive values will occur at times $T, 2T, 3T, \ldots nT$.

Spectral analysis works on the principle that any complex rhythm can be analysed as the sum of simple rhythms of different wavelengths. A commonly used form of spectral analysis is **Fourier analysis**, in which a complex rhythm is analysed as the sum of an infinite number of sinusoidal oscillations of different wavelengths. A **spectrogram**, or spectral density plot, is a plot of how oscillations of different wavelengths contribute to the total variance of the time series. Spectral analysis is particularly effective at analysing complex time series composed of multiple rhythms of different wavelengths. It can detect the presence of one or more ultradian rhythms superimposed over a dominant circadian rhythm or analyse the complex brain-wave patterns of an electroencephalogram (EEG) into simpler component waves of different frequencies.

Multiple regression analysis is another way of calculating how rhythms of different wavelengths contribute to the total variance in the data. Multiple regression allows simple trends in the data resulting from non-stationarity (such as linear changes in amplitude over time) to be removed. When combined with spectral analysis, this method is effective at analysing complex rhythms containing multiple periodicities and trends.

Bear in mind that even if spectral analysis or multiple regression analysis does indicate the presence of a periodicity in behavioural data, the amplitude of the variation might be small and the rhythm need have little biological significance.

Choice tests

The design of choice tests was discussed in Chapter 8. Here we consider ways in which the data may be analysed. If each subject's reaction to a stimulus is recorded as an all-or-nothing response – for example, 'approaches' or 'does not approach' the stimulus – the simplest method of presenting the results is in terms of how many subjects responded to each stimulus. However, this measure of response is relatively crude. If quantitative measures were obtained of how much or how many times each subject responded to the stimuli, which they should have been if at all possible, this additional information should be used.

Absolute differences in responsiveness are obtained by subtracting the response to stimulus R from the response to stimulus S. This provides a single score for each subject. **Response ratios** can be calculated in a variety of ways, a particularly useful form being:

Responses to R/(Responses to R + Responses to S)

If the subject has always responded to R and never to S, its score will be 1.0. Conversely, if it has responded to S but not R, its score will be 0. The chance level of response is 0.5. If more than two stimuli have been used, the divisor is the total number of responses to all the stimuli and the chance score is 1.0 divided by the number of stimuli used (for example, 0.25 in the case of four stimuli).

Subjects that fail to respond to either stimulus (*non-responders*) can cause problems in the analysis stage since calculating their scores involves dividing by zero. This problem can be dealt with by either systematically excluding the scores of all non-responders, or arbitrarily assigning a chance-level score (0.5 for a two-choice test) to all non-responders.

For the purposes of statistical analysis, each subject provides one value, which is the difference between its response ratio ($R/R + S$) and the chance level of 0.5 for a two-stimulus test. These difference scores are analysed with a matched-pairs test, as in the case of absolute differences. Results based on absolute rather than response ratio scores will be greatly affected by those individuals that have responded strongly in the tests. By contrast, the ratio method is more sensitive to variation within individuals, even when considerable variation is found between individuals. Indeed, it is logically possible for the two methods to generate contradictory results if the high-responding individuals have a preference for stimulus R and the low-responders a preference for S. For this reason, it is wise to analyse results both ways and, if the two methods generate different conclusions, consider why the differences have arisen. The apparent contradiction may itself be revealing.

Social behaviour

This section covers analysis of some of the commonly observed associations and interactions between individuals.

Indices of association

In studies of social behaviour it may be helpful to obtain an index of the extent to which two individuals, G and H, associate with each other. Any measure of association between the two must not only take account of the number of separate occasions that G and H are seen together, but also the number of separate occasions that G is seen on its own and the number of separate occasions that H is seen on its own. A simple and straightforward measure of association is calculated as follows:

$$\text{Index of association} = N_{GH}/(N_G + N_H + N_{GH})$$

where N_{GH} is the number of occasions G and H are seen together; N_G is the number of occasions G is seen without H; and N_H is the number of occasions H is seen without G. This index has the merit that all scores are distributed between 0 (no association) and 1.0 (complete association). A score of 0.5 means that the two individuals are seen apart as often as they are seen together. The information can be presented in a diagram, called a **sociogram**, depicting the extent of the associations between individuals. Individuals are connected to one another by lines, with the thickness of each line representing the degree of association between them (see Fig. 10.4).

For most indices of association, the chance level of association is not easily calculated and can only be estimated roughly from the likely constraints on the random movements of two independent bodies. Clearly, the larger the area in which individuals can move, the less likely they are to meet by chance. In the absence of a known chance level of association,

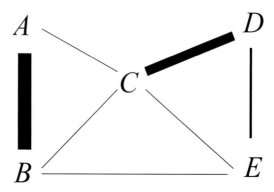

Figure 10.4 A hypothetical sociogram of strengths of association between individuals represented by the thickness of the lines between them. *A* and *B* associated strongly with each other as did *C* and *D*. *A* never associated with *D*, and *E* and *B* never associated with *D*.

the use of an index of association is restricted to making comparisons between, say, the two sexes or different age classes.

One way of estimating the chance encounter rates between pairs of animals is based on the random gas model, assuming that randomly interacting animals behave like molecules of gas. This method was used, for example, to assess whether the apparent pairings between certain male and female nocturnal lemurs were owing to chance meetings or whether they were indicative of deliberate associations (Schülke & Kappeler, 2003). As encounter rates in all pairs of lemurs were significantly greater than the predicted values, it appeared that animals were associating with each other more frequently than expected by chance. If encounter rates were lower than predicted, this would indicate active avoidance between individuals. Encounter rates similar to the predicted values would imply that animals lack interest in each other.

Maintenance of proximity

Two individuals, such as a mother and her offspring, may spend a lot of time together. One important measure of their relationship is the extent to which their proximity is owing to the movements of one member of the dyad rather than the other. This is measured by counting the occasions on which one member of the dyad approaches or leaves the other, while the other member remains still. (Strictly speaking, both individuals could be moving. The more precise question is: which individual's movements produced a crossing of an imaginary circle of specified radius around the other individual?) The number of occasions when the pair came together or were separated as the result of each individual's movements can then be obtained. A measure of the extent to which individual M has been responsible for maintaining proximity between itself and individual O can be calculated as follows:

$$M\text{'s responsibility for proximity} = U_M/(U_M + U_O) - S_M/(S_M + S_O)$$

where U_M is the number of occasions when the pair were united by M's movements; U_O is the number of occasions when the pair were united by O's movements; S_M is the number of occasions when the pair were separated by M's movements; and S_O is the number of occasions when the pair were separated by O's movements. The index ranges from -1.0 (O totally responsible for maintaining proximity) to $+1.0$ (M totally responsible). A zero value indicates that M and O were equally responsible for maintaining proximity. As with measures of association, the main value of such an index lies in making comparisons; for instance, the index has been used to show how the role of a rhesus monkey mother in maintaining contact with her offspring steadily declines as the offspring gets older (Hinde & Atkinson, 1970).

Dominance hierarchies

In many social species, the relationships between pairs of individuals are asymmetrical. One individual will consistently supplant the other when they compete for a valued resource such as food, shelter or a mate, or may simply cause it to move away when they meet. If the numbers of such occurrences are recorded for every pair in the group, it often becomes apparent that one individual tends to supplant all the other animals, whereas another is supplanted by all others. In between the top- and bottom-ranking individuals are animals that supplant some but are supplanted by others. The overall arrangement of dominant and subordinate individuals in the group is referred to as a dominance hierarchy. Some of the difficulties of interpreting such hierarchies or taking them too literally are described in the end of Chapter 11.

To derive a dominance hierarchy from observations of the interactions between individuals, the numbers of supplants between pairs are arranged in a matrix. The order is then arranged so that the individual that is never supplanted is at the top and the one that is always supplanted is at the bottom. The other animals are re-arranged in order until the minimum number of supplants appears on the left-hand side of the diagonal. The final order of a hypothetical case is shown in Fig. 10.5, where the individuals have been arranged into a dominance hierarchy.

If, as in this example, all individuals in the group can be arranged in strict order of dominance (*C* dominates *A*, *A* dominates *D*, *D* dominates *E* and *E* dominates *B*), then the dominance hierarchy is said to be **linear**. In reality, however, few dominance hierarchies are perfectly linear. Sometimes dominance reversals may occur, when a subordinate wins an encounter with a normally dominant individual. Moreover, for a hierarchy to be perfectly linear all dyadic relationships must be asymmetric, whereas in some groups two or more individuals may have equal status. Furthermore, in a perfectly linear hierarchy all possible triadic relationships must be **transitive**, which means that if *A* dominates *B*, and *B* dominates *C*, then *A* must also dominate *C*.

The procedure of arranging individuals to form a dominance hierarchy, as shown in Fig. 10.5, can be deceptive because there is a surprisingly high probability that a set of data can be arranged to form an apparent linear dominance hierarchy when none exists in reality. This is especially likely when the group is small and the observer, having no independent knowledge of the animals' relationships, *assumes* the existence of a linear hierarchy and juggles the data until the best hierarchy is obtained.

In its simplest form, as described above, the index of dominance status that is assigned to each individual is its rank in the hierarchy. Thus, dominance is measured on an ordinal (ranking) scale, which means in turn that the magnitude of the difference in dominance status between two

Number of occasions when individual was supplanted

		C	A	D	E	B
	C	–	22	8	18	11
	A	0	–	29	11	21
	D	0	0	–	6	11
	E	0	0	0	–	2
	B	0	0	0	0	–

Number of occasions when individual supplanted another

Figure 10.5 A hypothetical example of how five individuals (*A, B, C, D & E*) are rank-ordered on the basis of pair-wise interactions in which one supplanted the other. The matrix has been arranged so that the individual (*C*) that supplanted all the others is put first, the individual (*A*) that supplanted all others except *C* is put second, and so on down to *B* that supplanted none of the others. This rank order is the dominance hierarchy and, in this case, is perfectly linear.

individuals cannot be quantified. However, dominance can be measured on an interval scale, using a method of paired comparisons (Boyd & Silk, 1983). With this index of dominance, the difference in dominance between two individuals can be quantified and tested for statistical significance. This method is particularly useful for describing dominance hierarchies that are not highly linear and it can also be used for analysing asymmetric interactions that involve an actor and a recipient, such as grooming or food-sharing.

Besides linearity, another property of a dominance hierarchy is its **steepness** (De Vries *et al.*, 2006). Steepness measures the degree to which individuals differ from each other in winning dominance encounters. Linearity and steepness are complementary measures to characterise a dominance hierarchy.

The variety of techniques in use for analysing dominance hierarchies has proliferated, while the rationale for selecting a particular method has not always been adequately justified. A note of caution is therefore justified. Bayly *et al.* (2006) applied eight popular indices to the same set of data relating to interactions between male domestic fowl, the animals in which social hierarchies were first described historically. Overall agreement among methods was good when groups of birds had perfectly linear hierarchies, but results diverged when the social structure was more complex with, for example, intransitive triads or reversals. Choice of a particular index of hierarchy can therefore affect how social interactions are interpreted. We would advise you to explore different techniques when analysing your own data.

Summary

The temporal structure of behaviour can be examined in a variety of ways. Analysis of bout length shows whether a particular element of behaviour tends to be clumped in time and gives the duration of the clumps. Sequence analysis shows whether particular elements occur in a regular order. Rhythm analysis shows whether or not particular elements recur with regular periodicity. The results of choice tests may be analysed by comparing absolute levels of responsiveness or the relative preferences may be expressed as ratios. Indices of social behaviour include the extent to which individuals associate with each other, the extent to which one individual is responsible for maintaining contact, and dominance hierarchies.

11

Interpreting and presenting findings

Data can always be interpreted in a variety of ways. As you think about your findings, you would be wise to be sceptical, at least initially, about your own preferred explanation. Colleagues can usually be counted on to help you in considering alternative ways of accounting for your results. Many of the problems that arise in the interpretation of data can, however, be avoided if the research is carefully designed. We have discussed some of these issues in Chapters 7 and 8 – for example: are the measurements independent of each other? Is generalisation to another group of subjects limited because of the design? Have the results been affected by the order in which the treatments were presented? Did the observer influence the subjects in some way? Did you unwittingly select the data that fitted a preconceived idea or bias?

In well-designed research these issues will have been considered in advance and appropriate precautions taken. However, problems of interpretation that could have been foreseen often arise through oversight or the sheer practical difficulties of avoiding them. It is often hard, for example, to conduct an experiment 'blind'. The solution is not to ignore the problem, but rather to acknowledge it. You should be alert to potential difficulties and be honest about how these possibilities might affect interpretation. In this chapter we mention some additional issues that might affect how data are interpreted and discuss the presentation of results to a wider audience.

Floor and ceiling effects

Two groups of subjects may appear not to differ, when in reality they do, because all the scores are clumped at one or other end of the possible range of values. Genuine differences will be obscured if all or most subjects produce the minimum possible score (a **floor effect**) or the maximum possible score (a **ceiling effect**). For instance, a test of mathematical ability involving multiplication by two is unlikely to reveal differences between human adults because most people will answer all the questions correctly (a ceiling effect). A more difficult test is more likely to reveal differences, but a test that was too hard would result in most people answering none of the questions correctly (a floor effect). Although this point may seem obvious, it is often overlooked as a possible explanation when negative results are obtained. Floor and ceiling effects apply to correlations as well as differences, since two measures will appear to be uncorrelated if either set of scores is clumped at one end of its range of measurement.

Exploratory analysis of preliminary data should reveal whether floor or ceiling effects are likely to be a problem. If they are, matters may be improved by choosing a different, related measure that produces a broader spread of scores. For example, a successive choice test that simply measures whether or not two objects are approached may seem to show that both objects are equally attractive. However, a more sensitive measure of preference, such as latency to approach, may reveal clearer differences in the attractiveness of the two objects.

Assessing significance

Biological significance and statistical significance (see Chapter 9) are frequently confused when interpreting differences between groups or correlations between measures, especially when the difference or the correlation is small but statistically significant. Even when an effect is paltry, the finding can still be highly statistically significant if the sample size is big enough. With a sample size of 100, for example, a correlation of 0.20 is statistically significant ($p < 0.05$, two-tailed), even though a correlation of this size represents an extremely weak association ($r^2 = 0.04$; i.e. only 4% of the variation in one measure is accounted for statistically by variation in the other).

Table 11.1 *Informal phrases used to interpret correlation coefficients of different sizes (from Sprinthall, 2003).*

Value of correlation coefficient (r)	Informal interpretation
<0.2	Slight; almost negligible relationship
0.2–0.4	Low correlation; definite but small relationship
0.4–0.7	Moderate correlation; substantial relationship
0.7–0.9	High correlation; marked relationship
0.9–1.0	Very high correlation; very dependable relationship

Informal phrases can be used to interpret statistically significant correlation coefficients of various sizes (see Table 11.1).

The verbal tags attached to the correlation coefficients are subjective and they apply only to statistically significant correlations; their value lies in emphasising that statistically significant correlations may represent associations that are so weak as to be negligible under most circumstances. A related point is that, with the relatively small sample sizes used in some behavioural studies, correlation coefficients may provide fairly unreliable measures of the population values and may be difficult to replicate. Having issued this warning, we readily accept that small effects can be important in large-scale medical trials where a tiny but statistically significant association might mean the difference between life and death for a small segment of the population; moreover, over evolutionary time scales small effects can accumulate to generate large changes in phenotype.

While it is obviously essential to distinguish clearly between statistically significant and non-significant results, the level of significance by itself provides little useful information. For this reason we strongly recommend that the stated results should also include explicit information on the effect sizes (i.e. the magnitude of the differences between groups or the strength of the correlations between measures). When quoting a statistically significant difference between two groups, you should specify the actual mean scores, their associated measures of variation and the sample sizes. Simply stating that, say, 'the mean score for the experimental group was greater than that for the controls ($p < 0.05$)' is insufficient, as it gives no idea of how large this difference was. Similarly, when quoting a statistically significant correlation, state the actual correlation coefficient and sample size (or degrees of freedom), not just the level of significance. Ideally, results should also be accompanied by an indication of their precision – for example, in the form of a confidence interval.

Problems with correlations

The analysis of correlations between measures was described in Chapter 9 and without question provides a powerful tool in describing and understanding data. A number of pitfalls arise, however, when interpreting correlation coefficients.

Causality

A statistically significant correlation between two variables, *A* and *B*, can arise for one of three reasons: *A* causes *B*; *B* causes *A*; or *A* and *B* are independently related to a third variable, *C*. Problems frequently arise with this last case – that is, when the two measured variables are independently associated with an unknown third variable.

Linearity

A correlation refers to a linear relationship between two variables (strictly speaking such a relationship should be referred to as monotonic). Consequently, a correlation coefficient is meaningless if the two measures are associated in a non-linear (or non-monotonic) manner. Specifically, a lack of correlation between two measures does not demonstrate a lack of any association between them if they are non-linearly related. Figure 11.1 illustrates the case of two variables (*A* and *B*) which are strongly associated, but according to an inverse U-shaped relationship rather than a linear one. The correlation between *A* and *B* is zero, but it would obviously be wrong to infer that the measures are not systematically associated.

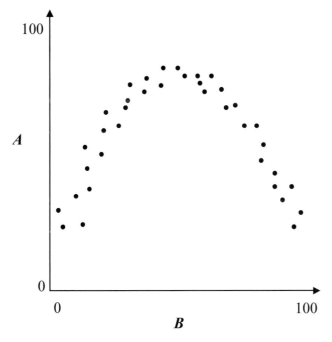

Figure 11.1 A scatter plot showing an inverse U-shaped relationship between two variables, *A* and *B*. The association between *A* and *B* is strong but non-linear. The correlation coefficient is zero and in this case it would be wrong to suppose therefore that the two variables were not associated.

It is essential to verify that two variables are not associated according to some non-linear relationship before calculating the correlation between them. The best precaution is simply to plot the data in a scatter diagram before calculating correlations.

Homogeneity

When interpreting a correlation it is usual to assume that the strength of association between the two variables applies across their entire range of variation – in other words, that the underlying population is homogeneous. Sometimes, however, this assumption is not justified because the strength

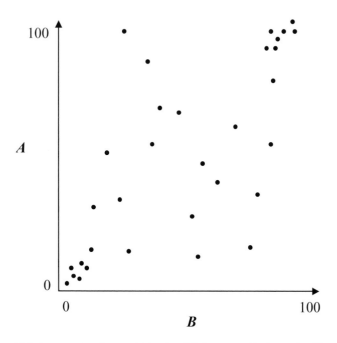

Figure 11.2 A scatter plot of two variables, *A* and *B*, that are positively correlated but only for extreme values of the two variables. The overall correlation is high ($r = +0.71$, d.f. $=$ 30, $p < 0.001$), yet the association between *A* and *B* is very weak for most of the range of scores.

of association is different for different sections of the population. Two variables (*A* and *B*) may show no association over most of their range, yet they can still be highly correlated if a strong association exists for extreme values only. This possibility is illustrated in Fig. 11.2.

The danger with this type of correlation is that it falsely implies a general association between two variables, when in fact the apparent association arises only from extreme cases. The association is therefore of limited validity. Equally, a low correlation can arise if two variables are positively correlated for one sub-group of the population (say, males) but negatively correlated for another (say, females). This effect is illustrated in Fig. 11.3.

Once again, the best precaution is to inspect the results visually on a scatter plot before calculating the correlation, since the presence of

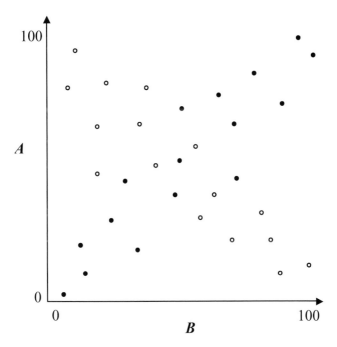

Figure 11.3 A scatter plot showing the association between two variables, *A* and *B*. The overall correlation is close to zero ($r = +0.08, p < 0.10$). However the sample is composed of two distinct sub-groups, represented by the open and filled circles. The open circles show a strong negative correlation ($r = -0.89, p < 0.001$) and the filled circles show a strong positive correlation ($r = +0.92, p < 0.001$).

distinct sub-groups or outliers within a sample will usually be apparent from a scatter plot.

Fishing expeditions

Correlating everything with everything else can be useful when preparing for factor analysis as described in Chapter 9 but it can lead to serious error. Calculating correlations for every possible pair-wise combination of variables is often referred to as 'fishing' when it reflects a failure to formulate clear hypotheses before the data were collected. In the absence of clear hypotheses or questions, the researcher resorts dubiously to analysing the data in every possible way in order to dredge up some statistically significant results, which are then explained by a post hoc 'hypothesis'.

The danger of fishing for correlations in this way is that if enough correlation coefficients are calculated, some statistically significant correlations are likely to arise by chance alone. Therefore any conclusions drawn from such analysis should take into account the *total* number of correlations that were calculated, both significant and non-significant. A significant correlation may seem less compelling if it turns out to be the only one out of, say, 45 correlations originally calculated. The issue is exactly the same as in multiple comparisons for which the Bonferonni correction is applied (see Chapter 9).

Systematically searching through data for interesting results is entirely admissible – indeed, recommended – provided it is seen solely as *exploratory* analysis, and not as confirmatory analysis or hypothesis-testing. Using a significant correlation as empirical proof of the post hoc 'hypothesis' which it gave rise to in the first place is dishonest.

Equivalence

Another problem with correlations can arise if all the variables were not measured with equal reliability and precision. If an independent variable (X_1) which is the real source of variation in behaviour (Y) has been measured badly, and it also affects another variable (X_2) which has been measured well, then multiple regression might indicate that X_2 accounted for the variation in Y better than X_1. You may then be led incorrectly to conclude that X_2 rather than X_1 was responsible for the differences in Y.

Treasuring your exceptions

Focusing solely on statistical significance can lead to an *under*-estimation of biological significance if, as is sometimes the case, an experimental treatment is more effective for some individuals in the sample than for others. If the group being studied is not homogeneous, then an experimental manipulation could produce an outcome that is not statistically significant for the group as a whole, yet still biologically highly significant for certain individuals in the group. For example, only a subset of humans is allergic to nuts.

Looking only for central tendencies in the data have led to serious scientific errors in the past. For instance, before the discovery of stratospheric ozone holes in the 1980s, statistical analysis of satellite data rejected the 'outliers' on the assumption that such measurements were unreliable. It was only when scientists working at one station in Antarctica repeatedly obtained low readings for atmospheric ozone that the processing mistake was discovered and the highly significant variation in the amount of the gas was properly appreciated (Farman, 1987).

Individual differences make it more difficult to draw valid conclusions about individuals from the characteristics of groups, since statements about the characteristics of a group may be untrue for some or all of the individuals within the group. The simplest precaution against being misled in this way is to plot the data for individuals separately before lumping them together. The odd exception to an overall trend may result from an error in recording or transcription. However, it may also be a finding to be treasured – as was sometimes the case in genetics when a mutation arose leading to abnormalities in structure, physiology or behaviour.

Prior knowledge and Bayes' theorem

When a body of knowledge already exists about prior occurrences of an event and another body of knowledge exists about the quality of measurement, the two sources can be combined. If, say, a bird-watcher spots a rare bird that has a 1% chance of occurring in that place, relative to other species with which it might be confused, and previous experience has shown that one in 20 of such positive identifications are wrong, what confidence should be placed in the bird-watcher's record? An eighteenth-century amateur mathematician called Bayes devised a theorem to help in such cases. Public confidence in a record, according to the theorem, is:

$$A \times B/((A \times B) + (1 - A) \times (1 - B))$$

Where *A* is the likelihood that the observer was right and *B* is the prior probability of occurrences of what was observed. Returning to the record of the rare bird and using the numbers from above:

Confidence in the bird-watcher's record $= (0.95 \times 0.01)/((0.95 \times 0.01) + (0.05 \times 0.99)) = 0.16$ or 16%

At first glance the result seems counter-intuitive and most people would assume that the confidence should be much greater than 16%. However, the calculation makes more obvious sense when considering the case of a bird that is so easy to identify that the false positive rate is zero. Then the confidence in the record is 100% even though the bird is rarely seen in that part of the world.

The use of Bayes' theorem tends to rely on educated guesses. 'Bayesians' interpret a probability statement as a degree of belief – confidence that a given statement is true. The theorem is most likely to be helpful in guiding judgements about whether or not a discovery is worth following up rather than providing a figure that could be included in the results section of a publication.

Modelling

Some biologists refer to a particular group of animals, such as a strain of mice, as being 'models' for humans with a particular condition or disease. In this section, though, we are concerned with models that represent explanations of things that happen in the real world. Statistical models that are used for analysing data are special cases (see Chapter 9). As you think about your own findings, you may be helped by special-purpose modelling software or, if you are good at computer programming, you may be able to write a program that will simulate the phenomenon you have been studying.

A model may attempt to explain how the nervous system generates behaviour, such as in the many examples of neural nets (Bateson & Horn, 1994; Enquist & Ghirlanda, 2005). The model may relate the performance of behavioural activities to a given end-point and represent the way an individual finds nutrients efficiently so that it survives to breed or help its kin to breed (Stephens & Krebs, 1986).

Formal models are usually smaller in scale and therefore simpler and easier to study than the phenomena they represent. The degree of simplification varies: some are deliberately over-simplified, almost like caricatures emphasising salient features of a process. Others are more elaborate, quasi-realistic and designed to include many of the likely factors influencing a process. Examples of quasi-realistic models are the extremely elaborate sets of equations used to forecast the weather, simulate traffic flow or understand the workings of the human heart. Even so, any model is at best a good approximation developed for a particular purpose.

The advantage of a well-formulated model is that it allows the essential features of a complex phenomenon to be explored systematically. Weaknesses in an argument expressed in words alone may be exposed. The model may encourage careful thought about what needs to be measured empirically. It may reveal outcomes that had not been expected at the outset. A big advantage of models is that they can allow hypothetical possibilities to be tested experimentally without using living subjects – an important consideration when concern about the use of animals in research is so great. In general, though, theoretical modelling is at its best when it is accompanied by empirical research on the real thing.

Models usually describe dynamic processes; this is achieved by specifying numerical values for a small number of terms called the parameters. Parameters are values that are needed to make the model work; they may be varied, but are kept constant for any one modelling event. (In Appendix 2 we mention the confusing and technically incorrect way in which 'parameter' is used as a synonym for 'variable'.) Giving sensible values to parameters requires suitable data or good background biological knowledge. The performance of a model may be poor if parameter values, about which little is known, have to be set arbitrarily.

Simple models can sometimes be generated by purely mathematical methods and expressed in an equation; the relations between the output of the model and the conditions that generate it are fully explicit. The reliance on computer simulations, rather than analytical mathematical equations, has the disadvantage that many separate runs of the computer may be needed to make comparisons between different sets of conditions, which in the analytical models are captured in a single mathematical formula.

However, computer-based simulations do have a powerful advantage, especially with the use of specially designed graphical displays, because they can show the evolution over time of the modelled process. Moreover, simulations are easy to run on computers (Mangel & Clark, 1988). The counter-balancing concern is that, in the absence of an intuitive understanding of how the model's assumptions and parameters relate to the precise conclusions, detailed quantitative output cannot always be trusted and must therefore be thoroughly tested against data. One standard approach in modelling is to test how robust its predictions are when the values of parameters are varied systematically – a process known as **sensitivity analysis**. In some models an element of chance variation is entered specifically into one or more of the variables. These models are called **stochastic**. When chance variation is not included the model is called **deterministic**. The introduction of chance into the model forces recognition of what is often uncertainty. Otherwise, deterministic models might be adequate, provided that the possibility of statistical fluctuation around the predicted outcomes is recognised.

In general, modelling should be seen as part of a process that continues from data analysis to designing new studies. Modelling may throw up possibilities that had not previously been considered. At a more mundane level it may stimulate empirical work to determine the numerical value of parameters whose values had been previously estimated. Above all, it may sharpen up the differences between alternative explanations for the data. A helpful guide to modelling is given by Kokko (2007), and Mangel (2006) provides a useful toolbox for the would-be theoretician.

Presentation of findings

Most scientific research is ultimately intended to be public, in the sense that researchers present their findings to colleagues by means of seminars, conference papers and posters, or publication in peer-reviewed journals. Much has been written about the best ways to do this. Use of videos in talks about behaviour can be especially effective, making points that are much more powerful than still pictures or oral descriptions. Good advice on presentations both oral and written is given in Lehner (1996) and Paterson (2001). Hailman and Strier (2006) provide a particularly helpful guide to the preparation of talks and written reports. Here we offer a brief outline of points to consider.

Introduction

Typically a written scientific paper will state why you did what you did, referring to previous published literature on the subject. Citing the work of others is an important part of setting the context for your own work. It also brings you into the community that is, by the nature of the enterprise, co-operative. Talking to colleagues and going to seminars are part of this process. Many aids are available on the web to help you, such as PubMed (www.ncbi.nih.gov) and the Web of Science (www.isiknowledge.com, but this requires you to be registered through an institution) and Google Scholar (www.google.com). Be aware that such search tools do not access records back to the beginning of time and important papers that relate to your own work may lie beyond their horizon. Not all scholarship can be conducted from a computer terminal and time spent in a good library can still be enormously valuable. When you know of an early important paper in your field of research, it may refer back to earlier literature. Furthermore the citation tools of the search engines will allow you to find other papers that refer to the one you already know. Yudkin (2006) offers good advice on how to read the scientific literature.

Methods

You should give a clear account of your methods so that someone else could repeat what you did. Descriptions might include the provenance of the animals and, if the study was done in the laboratory, how they were kept. You should include careful descriptions of your behavioural categories and how they were recorded and analysed.

Results

You should present your empirical findings without interpretation. Presenting quantitative information visually, in a way that is both informative and attractive, is both an art and a science. While it is too large a topic to cover adequately here, we urge you to think carefully about how to present your data. A graph, histogram, scatter plot or some other form of visual display is generally much more informative than densely packed tables of numbers. Moreover, as we have argued in Chapter 9, plotting data before analysing them statistically is a wise precaution against misinterpreting the statistics. We suggest the following general guidelines for plotting data.

Each axis of a graph should be clearly labelled with the name of the variable (e.g., average distance travelled, vocalisation rate, or amount of food consumed) and the units in which the variable is measured (e.g. km d^{-1}, min^{-1} or $g\,h^{-1}$). The type of measure used for behavioural categories should be clearly indicated: reciprocal units of time (s^{-1}, min^{-1}, d^{-1}, etc.) for frequency measures; and units of time (s, min, d, etc.) for duration and latency measures. Time-sampling measures and the proportion of time spent performing an activity are expressed as dimensionless scores between 0.0 and 1.0. Where a measure is expressed as the total number of occurrences, the period of time over which the events were counted should be indicated (e.g. 'per 30 min'). The dependent variable is conventionally drawn on the y-axis, with the independent variable on the x-axis.

Results are often presented in the form of multiple plots, all sharing the same x-axis but each with a different y-axis. This type of multiple plot can be misleading when casually inspected unless attention is drawn to the fact that the y-axis scales differ. It is doubly important in such cases that the axes are clearly labelled.

Graphs using a restricted range for the y-axis can also be misleading. The practice of using a restricted range along the y-axis, starting from a non-zero value, means that small variations in the dependent variable (y) are accentuated. It may be necessary and justifiable to focus on a particular part of the y-axis range, as though using a magnifying glass to inspect the data, since such inspection can reveal patterns in the data. But, unless it is also made clear that the *y*-axis shows only a selected portion of the range, the graph can be misleading.

When sample means (as opposed to the full range of raw scores) are plotted, it is often helpful to indicate the range or variation as well. This is usually shown by drawing a vertical bar through each point, the length of the bar representing the size of the standard error or sometimes the 95% confidence interval. The variability of scores about a median is often indicated with vertical bars denoting the interquartile range of the scores. If the sample sizes at different points on a graph are different, the sizes should be clearly indicated, usually in the caption.

Discussion

After presenting your results, you should discuss various ways in which they may be interpreted. While a risk-averse person may be over-cautious in interpreting data, you can also generalise too much. As an example, consider the interpretation of a dominance hierarchy. One common error is to over-generalise the meaning of a dominance hierarchy, by treating the dominance status of each individual as though it were a fixed and general characteristic of that individual. In fact, dominance relations are often fluid and capable of rapid change. Dominance relations sometimes have a geographical element, an individual's rank increasing towards the centre of its home range. In such cases, dominance reversals must be interpreted in the light of where the encounter occurs. Furthermore, a dominance hierarchy derived from one measure, such as competitive interactions over food, is not always the same as the hierarchy derived from a different measure, such as competition for mates.

Abstract

You will need to provide with your report a succinct summary of the research. A clear abstract is very important because, in most cases, it may be the only part of your paper that is read. It should provide the stimulus to an interested reader to go further and read the rest of the paper.

References

All literature that you have cited must be referred to properly. Annoyingly, scientific journals have not standardised the way that sources are printed or are referred to in the text. However, software tools such as Endnote (www.endnote.com) allow you to select the style that a particular journal uses.

Science and the public interest

Behavioural findings are often of interest to members of the public. That may be because such discoveries give them insight into their own and others' behaviour. Or, like discoveries about the universe, the public may find new work on behaviour intrinsically enjoyable – as is evident from the large audiences for TV programmes about the natural world. However, behavioural research may also have implications for the public interest in the sense that it potentially bears on matters such as eating habits, lifestyle, patient welfare, personal security and well-being, the state of human society and the conservation of the natural environment. As a researcher you have a social responsibility to form a realistic assessment of the potential implications of your work for society. You also have a responsibility to ensure the timely and appropriate communication to the public of your findings, if such communication is in the public interest.

Your work might bring substantial benefits to the public. But, if it is over-generalised or misreported, the consequences could be damaging. The unsubstantiated link between the MMR vaccine (for measles, mumps and rubella) and autism in children was an example. The misleading media reports in the United Kingdom led to many parents not having their children vaccinated and, as a consequence, local measles epidemics ensued. To help you meet your social responsibilities to the public, a checklist of questions to ask yourself is included in Appendix 4. If your work does have implications for the public interest, you may have to deal competently with the media and prepare what is called a 'lay summary' of your findings. This summary should be clear, concise and intelligible to somebody who has not been trained in science. Writing such lay summaries involves skills that many scientists lack. Dealing with the media also requires skills. Increasingly, funding bodies and universities are providing training courses to scientists to help them develop competence in public communication. Not everybody will want to acquire these skills, but the training is increasingly seen as a core element of continuing professional development. Our general point is that you should always be alert to the wider implications of your work, both positive and negative. Further discussion of these issues can be found in the Royal Society (2006) publication.

Honesty in research is crucial, both for the sake of your own reputation and the reputation of science as a whole. In the United States, the Office of Research Integrity is charged with overseeing the conduct of publicly funded biomedical and behavioural research. Among other things it promotes good practice, issues guidelines, monitors compliance and investigates cases of misconduct (http://ori.dhhs.gov/). Other countries also have established comparable bodies or are preparing to do so. As cases of faking evidence are revealed and public concern about scientific misconduct grows, every scientist needs to be aware of how much the whole enterprise depends on trust. At every stage of your research, therefore, you should stick firmly to the classical scientific virtues of honesty, scepticism and integrity.

Summary

Interpretation of data requires considerable self-criticism and thought. Findings may be influenced by floor or ceiling effects. High levels of statistical significance do not necessarily indicate large or interesting effects. Correlations between dependent variables raise questions about causality, linearity and homogeneity. Fishing for significant correlations among a large number of possibilities is a highly dubious practice. If all the variables in a study have not been measured with equal precision and reliability, conclusions about causality are led astray. When examining individual differences do not exclude outliers without careful thought, and treasure your exceptions. Bayes' theorem can indicate how much confidence should be given to the reliability of a finding. Theoretical modelling of the processes that gave rise to behaviour is often helpful in making sense of the data and in planning further work. Presentation of findings in whatever form is a crucial part of a research project. In doing so, consider whether the findings might benefit or damage the public interest and, at all times, maintain scrupulous honesty.

Appendix 1

Units of measurement

The SI system of units (Système International d'Unités) should be used for measurements. The SI system is completely *coherent*, which means that all derived units are formed by simple multiplication or division of base units without the need for any numerical factors or powers of ten. This distinguishes the SI system from earlier metric systems such as the centimetre–gramme–second (CGS) system, which it superseded. The SI system comprises nine base units, each of which is independently defined, and various other units which are derived by combining two or more base units. The base units, together with some of the more common derived units, are listed in Table A1.1. Some common non-SI units and their SI equivalents are shown in Table A1.2.

Conventions. Each unit is represented by a standard *unit symbol* (e.g. m, s, A, kg), which may be multiplied or divided by other unit symbols or numbers (e.g. 3 m, 0.12 kg m, 16.5 m s^{-2}). Unit symbols are algebraic symbols and follow the conventions of algebra. They are not abbreviations, and should never be followed by a full stop or an 's' (to denote plural). The names of units (e.g. metre, second, ampere) are all spelt with a lower case initial letter. Symbols for units named after a person start with an upper case letter (e.g. A for ampere, K for kelvin, Pa for pascal). When a measurement is used as an adjective, the number and unit should be joined by a hyphen (e.g. three-metre tube, 5-m distance, 9-s delay, 6-m^2 area, seventy-kilogram adult, 2-A current).

Standard prefixes. As shown in Table A1.3 standard prefixes are used to denote units multiplied by various powers of 10 (e.g. cm $= 0.01$ m, mA $= 0.001$ A, km $= 1000$ m). The prefix and unit symbol together should be treated as a single algebraic entity (e.g. 10 cm s$^{-1} = (10 \times 0.01$ m) s^{-1}).

Table A1.1 *The nine SI base units and some common derived units.*

Unit	Symbol	Quantity	Equivalent in base units
metre	m	length	BU[c]
kilogram	kg	mass	BU
second	s	time	BU
ampere	A	electric current	BU
kelvin	K	temperature[a]	BU
mole	mol	amount of substance	BU
candela	cd	luminous intensity	BU
radian	rad	angle	BU
steradian	sr	solid angle	BU
hertz	Hz	frequency[b]	s^{-1}
newton	N	force	$kg\,m\,s^{-2}$
joule	J	energy	$N\,m$
watt	W	power	$J\,s^{-1}$
pascal	Pa	pressure	$N\,m^{-2}$
coulomb	C	electric charge	$A\,s$
volt	V	potential difference	$J\,C^{-1}$
ohm	Ω	resistance	$V\,A^{-1}$
farad	F	capacitance	$C\,V^{-1}$

[a] In practice, temperature is usually measured in degrees Celsius (°C). One degree Celsius is exactly equal in magnitude to one kelvin, and 0 °C is approximately equal to 273 K. Use of the term centigrade in relation to temperature is, strictly speaking, incorrect, since a centigrade is a unit of angle. Note that the symbol for the unit of temperature (kelvin) is K, and not °K.

[b] Some authorities suggest that Hz should only be used to describe the frequency of periodic (regularly occurring) phenomena with a period of less than one second. When describing the rate of occurrence (frequency) of a behaviour pattern, it is normal practice to use the symbol s^{-1}.

[c] The abbreviation BU denotes base unit.

Table A1.2 *Some commonly used non-SI units.*

Unit	Symbol	Quantity	SI Equivalent
ångström†	Å	length	1.000×10^{-10} m
atmosphere†	atm	pressure	1.013×10^{5} N m^{-2}
calorie (thermochemical)†	cal	energy	4.184 J
Calorie (nutritional)†	Cal	energy	4.184 kJ
decibel[a]	dB	{sound intensity {sound pressure	–
degree Celsius†	°C	temperature	1.000 K
degree Fahrenheit	°F	temperature	0.556 K
hectare†	ha	area	1.000×10^{4} m^{2}
horsepower	hp	power	7.457×10^{2} W
inch†	in	length	2.54×10^{-2} m
kilogram-force	kgf	force	9.807 N
litre[b]	l	volume	1.000×10^{-3} m^{3}
millimetre of mercury	mmHg	pressure	1.333×10^{2} N m^{-2}

The above non-SI units are frequently encountered. Units marked with a dagger (†) are now defined in terms of SI units and the SI equivalent shown is an exact equivalent.

[a] The decibel (dB) is not strictly a *unit* of sound intensity, but denotes the ratio 10 $\log 10\{I/I'\}$, where I is the sound intensity (measured in W m^{-2}) and I' is a standard sound intensity (usually 10^{-2} W m^{-2}). It is also commonly used to denote sound pressure, again in terms of a ratio with respect to some defined reference level.

[b] The litre is actually defined in terms of a mass of water, and is not exactly equivalent to 1×10^{-3} m^{3}. To avoid confusion, most authorities recommend using dm^{3} instead.

Table A1.3 *Prefixes for SI units.*

T	tera	10^{12}
G	giga	10^{9}
k	kilo	10^{3}
h	hecto	10^{2}
da	deka	10
d	deci	10^{-1}
c	centi	10^{-2}
m	milli	10^{-3}
μ	micro	10^{-6}
n	nano	10^{-9}
p	pico	10^{-12}

Appendix 2

Some statistical terms

This list of terms is to start you off. A much fuller glossary of terms is given in Sprinthall (2003).

Degrees of freedom (d.f) gives the number of scores that are free to vary once certain restrictions have been placed on the data. To illustrate the point, when placing eggs into a standard egg box that takes six eggs (n), only one depression would be left for the sixth egg, allowing no 'freedom' about where it can be placed. Hence the degrees of freedom are $(n - 1) = 6 - 1 = 5$. The larger is the sample size, the greater is the number of degrees of freedom. In statistical analysis a degree of freedom is lost for every additional treatment that is added.

Effect size refers to the magnitude of an effect (or, in medical parlance, the 'clinical significance'). Effect size and statistical significance are quite separate matters and the level of statistical significance does not, as is often supposed, directly measure the magnitude or scientific importance of the observed result. A correlation or a difference can be very small in size yet highly statistically significant, provided the sample size is large enough (see Chapter 11 for further discussion).

Error: Measurement error is the combined error that results from inevitable imperfections and variability in the process of measurement. Measurement error may be either **random** (the scores are equally likely to be greater or less than their true values) or **systematic** (the scores are either consistently greater than or consistently less than their true values). **Sampling error** is the difference between a sample statistic (such as the sample mean) and the population parameter that is being estimated (such as the population mean). Sampling error arises because a finite sample rather than the whole population is measured. It can occur in any study where the entire population is not measured – in other words, in virtually every study ever conducted. Sampling error will diminish as the sample size is increased.

Factor is a term applied in the analysis of variance (ANOVA) and is an independent or grouping variable that must be either discontinuous (e.g. low/medium/high) or categorical (e.g. male or female).

Level, also used in analysis of variance, refers to a particular treatment or defined condition for the factor that is set or chosen by the researcher such as the degree of enrichment of housing conditions (e.g. low, medium or high).

Level of statistical significance (or **alpha level**) is the probability of obtaining the observed result, or one more extreme, if the null hypothesis were true. It is the probability that the observed effect, such as a difference or correlation, arose by chance alone (through sampling error) and that there is no real underlying effect. If this probability falls below a pre-determined critical level of significance, which is usually set at 0.05 (or 5%), then the null hypothesis is rejected. Thus, with the critical level of significance set at 0.05 the probability of rejecting the null hypothesis when it is in fact true (i.e. scoring a 'false positive' by concluding falsely that there is an effect) is less than one in 20.

Normal distribution (also called Gaussian distribution) refers to the frequencies of values for a variable which are approximately bell-shaped. If a variable is normally distributed, approximately two-thirds (68%) of all scores will fall within ±1 standard deviation of the mean, roughly 95% within ±2 standard deviations, and virtually all scores (99.7%) within ±3 standard deviations of the mean.

Null hypothesis is the baseline assumption made when testing a hypothesis, against which the outcome is compared. The null hypothesis is usually that no effect will be found and the results are owing to chance; for example, that no significant difference or no significant correlation exists between two sets of measurements.

One-tailed and two-tailed tests. If a prior prediction is made about the *direction* of an effect – for example, that the mean score for the experimental animals is greater than that for the control group – then the test is said to be **one-tailed.** Alternatively, if no direction is specified in advance (the prediction is simply that the scores are different), then the test is **two-tailed.** For instance, suppose you wished to test whether the weights of males and females differed significantly. If the initial hypothesis was simply that the sexes differ, then the test is two-tailed. If, however, existing knowledge or theory predicts the direction of the difference (for example, that males are heavier than females), then the test is one-tailed. For a one-tailed test to be used the prediction must have been made in advance of obtaining any results. Changing from a two-tailed test to a one-tailed test once the results are known is downright dishonest.

Parameters and variables. In general usage, a **parameter** is any quantity which is *constant* in the case being considered, but which may vary between different cases. 'Parameter' is sometimes used incorrectly as a synonym for *variable*, when referring to a behavioural category or some other measure. A **variable** may be **independent** of measured outcomes, such as a range of different treatments, or it may be **dependent** on the treatments. In statistical terminology, a parameter is a numerical characteristic, such as a mean or variance, describing the entire population. Parameters are usually estimated from samples rather than measured directly. Parametric statistical tests are so called because they make various assumptions about population parameters such as frequency distributions. They almost always assume normal, bell-shaped distributions.

Population is the entire set of items or individuals (animals, vocalisations, testosterone levels, etc.) under consideration, having at least one characteristic in common, and about which statistical inferences are to be made. The population is distinguished from a **sample**, which is the particular group or subset of entities selected from the population for measurement. Measurements are made on a sample (for example, ten mice from a laboratory colony of 100) and these measurements are then used to draw statistical inferences about the population as a whole (for example, the whole colony, different strains of mice living under comparable conditions, or even all mammals). It can often be difficult deciding how far a given set of results can be generalised – in other words, specifying the population to which the sample results refer. The temptation is usually to over-generalise results; for example, by assuming that the results of a laboratory experiment on learning in rats can be applied directly to learning in all animals, including humans.

Power of a statistical test is the probability of rejecting the null hypothesis when it is false; in other words, the probability of finding a real effect. The greater the power of a test, the more likely it is that a real effect, such as a difference or a correlation, will be detected. If the probability of a Type I error is fixed (e.g. at 5%), the statistical power can be increased by increasing the sample size (n) or by improving the research design; for example, by decreasing measurement error.

Samples. A **random sample** is a sample selected in such a way that each element or individual in the entire population has an equal chance of being selected for the sample. Truly random sampling is often referred to, but seldom achieved in practice. A **representative sample** is one that has the same broad characteristics as the population – for example, the same distribution of ages and weights, or the same sex ratio. A sufficiently large random sample will, on average, also be a representative sample. A **haphazard sample** is one where the sample is chosen according to an arbitrary criterion such as availability or visibility. Choosing randomly is notoriously difficult. In many studies, supposedly 'random' samples are actually more like haphazard samples and may not be fully representative of the population.

Type I and Type II errors. Rejecting the null hypothesis when it is in fact true (scoring a false positive) is referred to as a **Type I error.** Accepting the null hypothesis when it is false (scoring a false negative by failing to detect a real effect) is called a **Type II error.**

Standard deviation is a measure of variability in a normally distributed population, indicating how far all scores deviate from the mean value. It is the square root of the variance.

Standard error of the mean (SEM) is estimated by dividing the standard deviation of the sample by the square root of the sample size (n). Note that, as the size of a sample from a given population is increased, the SEM will decline. However, the benefits of reducing the measure of variability are reduced progressively with increased sample size.

Appendix 3

Advice on statistics textbooks

This edition of *Measuring Behaviour* does not contain the long annotated bibliography found in the first two editions. That bibliography contained references to the most important publications relating to the development of the methodology used in the direct observation of behaviour and has historical interest. For those who would like to consult it, you can visit the annotated bibliography of the second edition on the following website: www.cus.cam.ac.uk/~ppgb/.

The brief annotated bibliography contained in this edition relates solely to statistics books that we have found useful. It is far from comprehensive but should provide a helpful entry into a large and ever-growing library of books on the subject.

Annotated bibliography

Agresti, A. (2002). *Categorical Data Analysis*. 2nd edition. New York: Wiley-Interscience.
  A clear and up-to-date text on the treatment of categorical or nominal data.
Dancey, C. P. & Reidy, J. (2004). *Statistics without Maths for Psychology*. 3rd edition, London: Pearson.
  An excellent book for those who fear statistics because of the mathematical content. It does exactly what it says it does on the cover.
Fowler, J., Cohen, L. & Jarvis, P. 1998. *Practical Statistics for Field Biology*. 2nd edition. Chichester: John Wiley and Sons.

Aimed at field biologists. Includes methods of multivariate analysis with descriptions of principal components analysis, cluster analysis and discriminant analysis.

Grafen, A. & Hails, R. (2002). *Modern Statistics for the Life Sciences.* Oxford: Oxford University Press.
Strongly recommended for those wishing to use the powerful and flexible General Linear Models when analysing their data.

Hair, J. F., Anderson, R. E., Tatham, R. L. & Black, W. C. (1998). *Multivariate Data Analysis.* 4th edition. Upper Saddle River, NJ: Prentice Hall.
Provides an in-depth review of techniques used in multivariate statistics, and the situations to which they can be applied.

Hawkins, D. (2005). *Biomeasurement.* Oxford: Oxford University Press.
A modern and sympathetic introduction to the design of experiments and analysis of biological data.

Mead, R. (1988). *The Design of Experiments. Statistical Principles for Practical Application.* Cambridge: Cambridge University Press.
Focuses on the thinking behind replication, randomisation and blocking. However, it requires considerable knowledge of statistics.

Ruxton, D. G. & Colegrave, N. (2003). *Experimental Design for the Life Sciences.* Oxford: Oxford University Press.
Modern treatment of an important subject.

Siegel, S. & Castellan, N. J. (1988). *Nonparametric Statistics for the Behavioral Sciences.* 2nd edition. New York: McGraw-Hill.
A revised and co-authored edition of Siegel's original and unrivalled 1956 classic. Probably still the best single text on non-parametric methods; clear, concise, reliable and indispensable.

Sokal, R. R. & Rohlf, F. J. (1995). *Biometry: The Principles and Practice of Statistics in Biological Research.* 3rd edition. New York: W. H. Freeman.
Often referred to as the 'bible' by statistically minded biologists. Probably most used to check on details not covered by more elementary books.

Sprinthall, R. C. (2003). *Basic Statistical Analysis*. 7th edition. Reading, MA: Addison-Wesley.

Excellent for providing easily understood explanations of the basic concepts of statistics and experimental design. Witty and clear.

Zar, J. H. (1999). *Biostatistical Analysis*. 4th edition. London: Prentice Hall.

One of the best general texts on statistics. More accessible than Sokal & Rohlf. A fifth edition is expected in 2007.

Appendix 4

Checklist to consult before publication

Reprinted from *Science and the Public Interest*, published by the Royal Society of London in 2006.

1. What implications, if any, do your research results have for the public, for instance in terms of:
   - the eating or lifestyle habits of consumers;
   - the well-being of patients;
   - personal security or other issues affecting the well-being of individuals;
   - the state of human society in general;
   - the state of the environment; or
   - public policy?

2. Would the communication of your results be in the public interest, in terms of:
   - furthering the understanding of, and participation in, the debate of issues of the day;
   - facilitating accountability and transparency of researchers, their funders and their employers;
   - allowing individuals to understand how the results of research affect their lives and, in some cases, assist individuals in making informed decisions in light of the results; or
   - bringing to light information affecting public well-being and safety?

3. Do you need any advice to help you to decide whether communication of your research results would be in the public interest, and if so whom do you need to assist you?

4. Are there any reasons why disclosure of your research results might not be in the public interest, such as national security considerations?

5. Are there any other interests, such as commercial confidentiality, stock market regulations or intellectual property rights, competing with the public interest in terms of the communication of your results?

6. Are you able to provide the appropriate context for your research results, such as:
   - indicators of the accuracy of the results (e.g. statistical significance);
   - indicators of the integrity and credibility of the results;
   - information about the ethical conduct of the research;
   - indicators of uncertainty in the interpretation of results;
   - expressions of risk that are meaningful; and
   - comparison of the new results with public perceptions, 'accepted wisdom', previous results and official advice?

7. Do you need any advice to help you to provide appropriate context for your results, and if so whom do you need to assist you?

8. How might your results be used by other individuals or organisations, such as campaigners or policy-makers?

9. To what extent have your results and their context been subjected to a review of their accuracy, integrity and credibility, for instance through a peer-reviewed journal?

10. In terms of the public interest, when would it be best to communicate your results to the public?

11. In terms of the public interest, what would be the best way for you to communicate your results to the public?

12. If you are presenting results at a scientific conference, is it in the public interest for them to be communicated to the public at this stage?

13. Is there a regulatory body which you should contact about your results?

14. Do you need to provide a 'lay summary' of your results and their implications for the public?

15. Have you checked any materials prepared for the media about your results?

REFERENCES

Bakeman, R. & Quera, U. (1997). *Observing Interaction: An Introduction to Sequential Analysis.* 2nd edition. Cambridge: Cambridge University Press.

Barrett, L., Dunbar, R. & Lycett, J. (2002). *Human Evolutionary Psychology.* Princeton, NJ: Princeton University Press.

Bateson, M. (2004). Mechanisms of decision-making and the interpretation of choice tests. *Animal Welfare Supplement,* **13**, S115–20.

Bateson, P. (2005a). Ethics and behavioral biology. *Advances in the Study of Behavior,* **35**, 211–33.

Bateson, P. (2005b). The return of the whole organism. *Journal of Biosciences,* **30**, 31–9.

Bateson, P., Barker, D., Clutton-Brock, T. *et al.* (2004). Developmental plasticity and human health. *Nature,* **430**, 419–21.

Bateson, P. & Horn, G. (1994). Imprinting and recognition memory: a neural net model. *Animal Behaviour,* **48**, 695–715.

Bausell, R. B. & Li, Y.-F. (2002). *Power Analysis for Experimental Research.* Cambridge: Cambridge University Press.

Bayly, K. L., Evans, C. S. & Taylor, A. (2006). Measuring social structure: A comparison of eight dominance indices. *Behavioral Processes,* **73**, 1–12.

Bennett, A. T. D., Cuthill, I. C., Partridge, J. C. & Lunau, K. (1997). Ultraviolet plumage recognition. *Journal of Theoretical Biology,* **81**, 65–73.

Boyd, R. & Silk, J. B. (1983). A method for assigning cardinal dominance ranks. *Animal Behaviour,* **31**, 45–58.

Burkhardt, R. W. (2005). *Patterns of Behavior*. Chicago, IL: University of Chicago Press.

Burley, N. T. (2006). An eye for detail: selective sexual imprinting in zebra finches. *Evolution, 60*, 1076–85.

Carere, C. & Eens, M. (2005). Unravelling animal personalities: how and why individuals consistently differ. *Behaviour, 142*, 1149–431.

Cohen, J. & Medley, G. (2000). *Stop Working, Start Thinking*. Oxford: Bios Scientific.

Cruze, W. W. (1935). Maturation and learning in chicks. *Journal of Comparative Psychology, 19*, 371–408.

D'Eath, R. B. (1998). Can video images imitate real stimuli in animal behaviour experiments? *Biological Reviews, 73*, 267–92.

De Vries, H., Stevens, J. M. G. & Vervaecke, H. (2006). Measuring and testing the steepness of dominance hierarchies. *Animal Behaviour, 71*, 585–92.

Diamond, J. & Bond, A. B. (2004). Social play in kaka (*Nestor meridionalis*) with comparisons to kea (*Nestor notabilis*). *Behaviour, 141*, 777–98.

Dunn, J. & Plomin, R. (1990). *Separate Lives: Why Siblings are so Different*. New York: Basic Books.

Enquist, M. & Ghirlanda, S. (2005). *Neural Networks and Animal Behavior*. Princeton, NJ: Princeton University Press.

Farman, J. C. (1987). Recent measurements of total ozone at British Antarctic survey stations. *Philosophical Transactions of the Royal Society of London, 323*, 629–44.

Gluckman, P. & Hanson, M. (2004). *The Fetal Matrix*. Cambridge: Cambridge University Press.

Grafen, A. & Hails, R. (2002). *Modern Statistics for the Life Sciences*. Oxford: Oxford University Press.

Hailman, J. P. & Strier, K. B. (2006). *Planning, Proposing and Presenting Science Effectively*. 2nd edition. Cambridge: Cambridge University Press.

Hernández-Lloreda, M. V. (2006). The utility of generalizability theory in the study of animal behaviour. *Animal Behaviour, 71*, 983–8.

Hinde, R. A. & Atkinson, S. (1970). Assessing the roles of social partners in maintaining mutual proximity, as exemplified by mother–infant relations in rhesus monkeys. *Animal Behaviour,* **18**, 169–76.

House of Lords (2002). *Select Committee on Animals in Scientific Procedures. Vol. I – Report*. London: The Stationery Office.

Hunt, S., Cuthill, I. C., Swaddle, J. P., & Bennett, A. T. D. (1997). Ultraviolet vision and band colour preferences in female zebra finches *Taenioypygia guttata*. *Animal Behaviour,* **54**, 1383–92.

Jablonka, E. & Lamb, M. J. (2005). *Evolution in Four Dimensions*. Cambridge, MA: MIT Press.

Jammalamadaka, S. R. & SenGupta, A. (2001). *Topics in Circular Statistics*. Hackensack, NJ: World Scientific.

Kahng, S. & Iwata, B. A. (1998). Computerised systems for collecting real-time observational data. *Journal of Applied Behavior Analysis,* **31**, 253–61.

Kenward, R. E. (2000). *A Manual for Wildlife Radio Tagging*. 2nd edition. London: Academic Press.

Kokko, H. (2007). *Modelling for Field Biologists*. Cambridge: Cambridge University Press.

Kroodsma, D. E. (1989). Suggested experimental designs for song playbacks. *Animal Behaviour,* **37**, 600–9.

Lahti, D. C. & Lahti, A. R. (2002). How precise is egg discrimination in weaverbirds? *Animal Behaviour,* **63**, 1135–42.

Lehner, P. N. (1996). *Handbook of Ethological Methods*. 2nd edition. Cambridge: Cambridge University Press.

Magnusson, M. S. (2000). Discovering hidden time patterns in behavior: T-patterns and their detection. *Behavior Research Methods, Instruments and Computers,* **32**, 93–110.

Mangel, M. (2006). *The Theoretical Biologist's Toolbox*. Cambridge: Cambridge University Press.

Mangel, M. & Clark, C. W. (1988). *Dynamic Modeling in Behavioral Ecology*. Princeton, NJ: Princeton University Press.

Mead, R. (1988). *The Design of Experiments*. Cambridge: Cambridge University Press.

Milinski, M., *et al.* Griffiths, S., Wegner, K. M. (2005). Mate-choice decisions of stickleback females predictably modified by MHC peptide ligands. *Proceedings of the National Academy of Sciences USA,* **102**, 4414–18.

Nuffield Council on Bioethics (2004). *The Ethics of Research Involving Animals*. London: Nuffield Foundation.

Ord, T. J., Peters, R. A., Evans, C. S. & Taylor, A. J. (2002). Digital video playback and visual communication in lizards. *Animal Behaviour,* **63**, 879–90.

Ozer, D. J. & Benet-Martinez, V. (2006). Personality and the prediction of consequential outcomes. *Annual Review of Psychology,* **57**, 401–21.

Paterson, J. D. (2001). *Primate Behavior*. 2nd edition. Prospect Heights, IL: Waveland.

Real, L. A. (1994). *Behavioural Mechanisms in Evolutionary Ecology*. Chicago, IL: University of Chicago Press.

Royal Society (2006). *Science and the Public Interest*. London: Royal Society of London.

de Ruiter, J. R. (1986). The influence of group size on predator scanning and foraging behaviour of wedge-capped capuchin monkeys (*Cebus olivaceous*). *Behaviour,* **98**, 240–58.

Ruxton, D. G. & Colegrave, N. (2003). *Experimental Design for the Life Sciences*. Oxford: Oxford University Press.

Sanson, A., Hemphill, S. A. & Smart, D. (2002). Temperament and social development. In P. K. Smith & C. H. Hart (eds.), *Blackwell Handbook of Childhood Social Development* (pp. 97–116). Oxford: Blackwell.

Sapolsky, R. M. & Share, L. J. (1998). Technical Report. Darting terrestrial primates in the wild: A primer. *American Journal of Primatology,* **44**, 155–67.

Schülke, O. & Kappeler, P. M. (2003). So near and yet so far. Territorial pairs, but low cohesion between pair partners in a monogamous lemur, *Phaner frucifer*. *Animal Behaviour*, **65**, 331–43.

Semple, S. & McComb, K. (2000). Perception of female reproductive state from vocal cues in a mammal species. *Proceedings of the Royal Society of London B*, **267**, 707–12.

Siegel, S. & Castellan, N. J. (1988). *Nonparametric Statistics for the Behavioral Sciences*. New York: McGraw-Hill.

Smith, E. L., Evans, J. E. & Párraga, C. A. (2005). Myoclonus induced by cathode ray tube screens and low-frequency lighting in the European starling (*Sturnus vulgaris*). *Veterinary Record*, **157**, 148–50.

Sprinthall, R. C. (2003). *Basic Statistical Analysis*. 7th edition. Boston, MA: Allyn & Bacon.

Stephens, D. W., & Krebs, J. R. (1986). *Foraging Theory*. Princeton, NJ: Princeton University Press.

Tinbergen, N. (1963). On aims and methods of ethology. *Zeitschrift für Tierpsychologie*, **20**, 410–33.

Watts, J. M. (1998). Animats: Computer-simulated animals in behavioral research. *Journal of Animal Science*, **76**, 2596–604.

Wemelsfelder, F., Hunter, T. E. A., Mendl, M. T. & Lawrence, A. B. (2001). Assessing the 'whole animal': a free choice profiling approach. *Animal Behaviour*, **62**, 209–20.

Wiley, R. H. (2003). Is there an ideal behavioural experiment? *Animal Behaviour*, **66**, 585–8.

Yudkin, B. (2006). *Critical Reading*. London: Routledge.

Zar, J. H. (1999). *Biostatistical Analysis*. 4th edition. London: Prentice Hall.

Zinner, D., Hindahl, J. & Schwibbe, M. (1997). Effects of temporal sampling patterns of all occurrence recording in behavioural studies: many short sampling periods are better than a few long ones. *Ethology*, **103**, 236–46.

INDEX